The
Great
Human
Diasporas

THE
GREAT
HUMAN
DIASPORAS

The History of Diversity and Evolution

Luigi Luca Cavalli–Sforza and
Francesco Cavalli–Sforza

TRANSLATED FROM THE ITALIAN BY SARAH THORNE

Helix Books

PERSEUS BOOKS
Cambridge, Massachusetts

Many of the designations used by manufacturers and sellers to distinguish their products are claimed as trademarks. Where those designations appear in this book and Perseus Books was aware of a trademark claim, the designations have been printed in initial capital letters.

Library of Congress Cataloging-in-Publication Data

Cavalli-Sforza, L. L. (Luigi Luca),
   [Chi siamo. English]
   The great human diasporas: the history of diversity and evolution/
Luigi Luca Cavalli-Sforza and Francesco Cavalli-Sforza; translated
from the Italian by Sarah Thorne.
     p. cm.
   Includes bibliographical references and index.
   ISBN 0-201-40755-8 (hardcover)
   ISBN 0-201-44231-0 (paperback)
   1. Human evolution.   2. Race.   3. Ethnic groups.   4. Human
genetics.   I. Cavalli-Sforza, Francesco, 1950–   II. Title.
GN281.C3813   1995
573.2—dc20   94-26138
                                        CIP

This work was first published in Italy by Arnoldo Mondadori Editore Spa under the title *Chi Siamo: La Storia della Diversità Umana* © 1993 Arnoldo Mondadori Editore Spa, Milano.

English translation copyright © 1995 by Perseus Books Publishing, L.L.C.
Postscript © 1995 by Luigi Luca Cavalli-Sforza

Cover design by Lynne Reed
Cover photograph © George Holton/Photo Researchers
Text design by Janis Owens
Set in 10-point Galliard by Science Typographers, Inc.

Perseus Books is a member of the Perseus Books Group

EBC 11 12 13 14 15 -- 07 06 05 04 03

Find us on the World Wide Web at
http://www.perseusbooks.com

Dedicated to the women who gave us our mitochondria

Contents

	Preface	ix
	Acknowledgments	xiii
ONE	The Oldest Way of Life	1
TWO	Portraits from the Past	27
THREE	One Hundred Thousand Years	49
FOUR	Why Are We Different? The Theory of Evolution	74
FIVE	How Different Are We? The Genetic History of the Human Species	106
SIX	The Last Ten Thousand Years: The Great Trek of the Cultivators	126
SEVEN	The Tower of Babel	164
EIGHT	Cultural Legacies, Genetic Legacies	203
NINE	Race and Racism	227
TEN	Evolution and Progress	245
	Postscript	267
	Bibliography	285
	Index	289

Preface

Luca Cavalli-Sforza, my father, is a scientist. He has dedicated forty years' work to the evolution of the human races, using information gathered about our genetic heritage and studying other sciences: archaeology, linguistics, anthropology, history, demography, and statistics. This book describes the questions asked, ideas analyzed, observations made, data collected, and interpretations that have emerged from his attempt to clarify the nature of our past. It is not the scientific biography of a researcher; if Luca is a central figure in his sphere, it is because of the originality of his research and his way of drawing together contributions from different areas to make them work together.

It is curious, and relevant, that a book dealing with continuity and change is jointly written by a father and a son. Originally an interview book, somewhere along the way it became a first-person account. The narrator in the text is Luca. He tells of his efforts as a scientist who posed problems and searched for answers, and as a man who found ways to cooperate with people superficially as different from him as African pygmies or superficially as similar as scientists in other, mostly humanistic, disciplines. We pooled our efforts to make the style approachable and comprehensible. The reader needs no specific knowledge, and the few technical terms are explained before use. The material is organized in subsections, to facilitate reading and consultation.

I am by trade a film director, not a scientist. My job is to spin tales. While preparing this book, I found myself face to face with history, with the paths trodden by those few thousands or tens of thousands of people who, over a period of one hundred thousand years, colonized every corner of the earth. We number almost six billion today, a figure that looks set to double in the time of perhaps two generations. For me, this has been an important opportunity to uncover and reflect on our past. I hope it will be the same for the reader.

Chapter One of this book discusses the pygmies and the last remaining tribes of hunter-gatherers, those who still practice the lifestyle that characterized the whole species until ten thousand years ago (about 99 percent of our existence, that is). Chapters Two and Three describe what we know of the development of humans up to one hundred thousand years ago and then, with modern humans, until now. Chapter Four explores the theory of evolution, and the forces that have joined or divided living creatures over time. Chapters Five, Six, and Seven tell the story of the peoples colonizing the planet in the last one hundred thousand years, the slow, inexorable expansion of agriculture in the last ten thousand years; and the extraordinary diversification of languages that has accompanied the spread of humankind. Chapter Eight addresses our cultural and genetic heritage. Chapter Nine looks at the crucial topics of race and racism. Chapter Ten, at the genetic future of humans, genetic engineering, and current attempts to describe fully the inheritance of every human being (the Human Genome Project). Although preferable, it is not necessary to read these chapters in order. Chapter Four, however, which describes the theory of evolution, is recommended to those wishing to understand the logic underlying this book. Chapters One, Five, Six, Eight, and Nine in particular contain new research by Luca Cavalli-Sforza, presented for the first time in a book aimed at a nonspecialist audience.

My father's job and mine have at least one thing in common—the collective element. Scientific research, like the making of a film, involves teams of dozens of people. The ability to communicate and cooperate, to think up and carry out ideas, to pose the right questions, and to find the best solutions are essential for the success of both. Nearly all the research presented here has been done by several people working toward a common objective.

It is encouraging to find that the oldest traces of human activity indicate that from earliest times our ancestors worked together. Apparently the ability to cooperate has always been one of our most promising features.

I believe that making history is preferable to writing it. But the past has always fascinated me. It is extraordinary that today, thanks to progress in research techniques, we can often know far more about antiquity than even the most informed person of the time. The thousands of generations before us have left us the results of their actions and their biological composition, and to them we owe our existence. But with each new day on earth, everything depends on us, the living, and the example we give to those being born.

Our history is written in our genes and in our actions. We can do little about the former, but virtually everything about the latter, if we are free people. Never more so than now—at the end of the second millennium of the Christian era, in the fourteenth century of the Islamic hegira, and twenty-six centuries after the enlightenment of Buddha—have we been able to make the earth a garden or a desert, and life a pleasure or a torture.

We must remember that what unifies us outweighs what makes us different. Skin color and body shape, language and culture, are all that differentiate the peoples scattered across the earth. This variety, which testifies to our ability to accept change, adapt to new environments, and evolve new lifestyles, is the best guarantee of a future for the human race. The knowledge acquired about ourselves and described in this book clearly shows, however, that all this diversity, like the changing face of the sea or sky, is minute compared with the infinite legacy we human beings possess in common.

FRANCESCO CAVALLI-SFORZA

Acknowledgments

We have many debts of gratitude to a great many people to whom we also extend our warmest thanks. All of our family members read our manuscript and often proposed important improvements. The same was done by Giovanni Magni, Professor of Genetics at Milan University; Guido Pontecorvo, former Professor of Genetics at Glasgow University; and Marco Vigevani of Mondadori. The following people reviewed specific chapters for us. Giacomo Giacobini, Professor of Anatomy at Torino reviewed Chapters 2 and 3, and Paolo Ramat, Professor of Glottology at Pavia University reviewed Chapter 7. None of these people are responsible for any errors remaining in the text. André Langaney of the Musée de l'Homme in Paris helped us trace an important original photograph. Mark Seielstad patiently and carefully revised the English proofs. We are sorry it was not possible to include all his suggestions. The drawing of the pygmy reading the book *African Pygmies* (p. 25), is inspired by a photograph taken by Barry Hewlett in the Central African Republic. Perhaps we should mention that the pygmy doesn't really know how to read (L. C-S. has met only two adult pygmies out of thousands who knew how to read).

We would like to thank the publishers who gave us permission to use their copyrighted material: Princeton University Press for the figures of the main components of Europe which appear in Chapter 6, taken from the book *History and Geography of Human Genes*, first published in 1994. Figure 8 of Chapter 3 was taken, with modifications, from an article in the journal *Science*; figure 6 in Chapter 4 is inspired, with many adaptions, by an image from *Tous parents, tous différents*.

THE
GREAT
HUMAN
DIASPORAS

ONE

The Oldest Way of Life

I am not a hunter by nature. Years ago, however, I was invited to a hunting reserve in Austria and gave in to temptation. In the reserve woods stood a blind with some steps leading to a small balcony, where a rifle lay ready on some cushions. After I had been waiting for a short while, a fine roebuck wandered clearly into view in a patch of sunlight, about one hundred yards away. I am a fairly good shot, but being an inexperienced hunter, I did not know quite where to aim. I hit the splendid animal between the chest and belly, and luckily he must have died almost at once. A moment later, filled with an enormous sense of guilt, I watched with a heavy heart the gamekeeper perform the ritual celebration of the death of an animal—dipping a pine sprig in its blood, then putting it in my hat. I told myself I would never hunt again.

I did hunt again, but under very different circumstances. In the 1960s, I began researching the African pygmies, who live by hunting and gathering their food in the wild. My work did not bring me into direct contact with their hunting techniques, but I became curious to see these great professionals at work in the tropical forests. I knew that it was the pygmies who had supplied the first Portuguese merchants with nearly all the ivory that they traded with the rest of the world after the Portuguese settled Africa's Atlantic coast. Then, as now, the pygmies lived in the forest, generally far away from where the Portuguese dropped anchor, using local cultivators as their intermediaries.

Pygmies hunt their own special way, without rifles. It is well known that they are small, and it is ironic that the world's smallest people hunt its largest animal. With enormous bravery, pygmy hunters wait for the elephant to charge and then they spear it, dodging out of the way at the very last moment. Alternatively, they wound their prey sideways in the haunches or stomach, or cut the leg tendons.

I did not go elephant hunting with the pygmies. The hunting ground was at least four or five days' march away, through the forest, where temperatures are close to 100° Fahrenheit and humidity nears 100 percent. One cultivator who witnessed an episode told me that at the crucial moment he hid behind a tree, overwhelmed with fear at the sight of the huge animal charging toward the pygmy *tuma* (the prestigious title accorded to experts in elephant hunting).

Hunting with the Pygmies

Pygmy society is without hierarchy. The "leader," who has no actual authority, is merely a reference point for outsiders. I asked the leader of a camp if a colleague and I could go net-hunting with his group. He spoke at length with the others, clearly taking into account that we had brought plenty of food and cigarettes as gifts.

The settlement comprised nine or perhaps ten families: seven agreed that we could accompany them. A net hunt—the type normally practiced by most pygmies—requires at least seven nets to form a wide enough circle. Each family usually has just one net, about forty yards long, made out of rope taken from the bark of certain trees.

We left the following morning, and set up camp in the forest a few hours later. The women spent two or three hours building huts in the shape of an elongated sphere, long enough for a pygmy to lie down in, with an entrance so small that you have to wriggle through on your stomach. The bare outer structure of entwined branches is covered by large leaves, making it completely waterproof. The bed is made of thin trunks laid lengthwise. Two young pygmies had no wives to build their hut, so they slept on a bed of branches in the open, huddled together against the cold. My colleague and I used our camp beds with mosquito nets. In the night it rained, and we donned our raincoats and propped our beds up under a tree to prevent them from getting too wet. The rain didn't last long (it was the dry season), so we were able to go back to sleep.

The next day we set out hunting, together with the women and smaller children, who went looking for birds and turtles. The dense tropical forest, with its thirty to forty-yard-high trees, grows a leaf cover thick enough to block out the sun, leaving the forest sunk in a deep shade. The only vegetation is the occasional bush or plant, which is extremely green because of the lack of light. The ground is strewn with fallen tree trunks and roots.

The men lay out their yard-high nets in a rough circle, hanging them from low branches to keep them from being obvious to animals. Each man remains silent and unseen by the others until the signal is given that the circle is complete. The hunt begins: three or four pygmies move toward the center with their spears, making noises to frighten the prey; the others, including the women, stay with the nets to grab the animals, mostly gazelles and antelopes, which, when caught, try to escape. They run away very fast, which makes them hard to seize. Visibility in the thick forest is limited to a few yards, so it is rare to see these encounters; shouts and struggling noises are heard until the prey either is captured or escapes. The hunt lasts forty or fifty minutes, and then the group moves about half a mile and starts again.

We went on like this all day, without much success. Between one netting and another, the pygmies tried to change their luck with magic words, spitting on the nets, sometimes enticing the animals with songs or hurling insults at them. At one point a big animal—an antelope—was taken. We knew because in the middle of all the confusion we suddenly heard a loud laugh, clearly an expression of great joy.

The meat was divided among all the members of the village, although some of the best parts belonged by right to those who had taken the animals. For the pygmies, hunting is a job, a necessity, but also a pleasure. Like the game of poker, it entails all the uncertainties of luck but requires skill and experience. The stake is to eat or go hungry. The pygmies have developed a great understanding of animal behavior, which allows them to hunt difficult animals such as the anteater and large ones such as the elephant. They are completely different from other hunters of our time, who hunt as a hobby and spend hours shut up in river blinds waiting for duck, perhaps risking being shot by another hunter, but never going hungry. The pygmies love their life profoundly. To uproot them is difficult—you have to destroy the forest to do so. That is exactly what has been happening over the last two thousand years, and it continues at an astounding rate, in a kind of planetary annihilation. But as long as large untouched areas of tropical forest survive in Africa, pygmies will be hunting in it. Their proverbial skill was proved to us when we lent one a rifle and four cartridges: he returned that evening with three dead animals and one cartridge, which had misfired.

The forest is full of a thousand pleasures and delicacies. For example, I fancied trying wild bee honey. A pygmy told me he knew of a hive three hours' journey away (and thirty yards up a tree!). I promised him a reward and he returned later with a dark-colored hive, full of very strong honey, some of which we mixed with whiskey.

Each spring there is a celebration of the arrival of the caterpillars. The forest fills with the apparently very tasty butterfly caterpillar. I was not in the area at the time. This is when the pygmies take the settled farmers into the forest and without knowing help them store up on protein, precious in the farmers' mostly carbohydrate diet.

A Geneticist-Witch

Genetics is the science of heredity. It is the key to biology because it deals with the mechanisms allowing the reproduction of living creatures, the working and transfer of hereditary material, and the differences between individuals, as well as biological evolution. These are the fundamental characteristics distinguishing living organisms from inanimate matter.

I am a geneticist. For the last forty years, much of my life has been dedicated to studying the evolution of human peoples. In 1966, while teaching genetics at Pavia University, I became convinced of the importance of organizing a scientific expedition to the pygmies of Africa's tropical forests. Why the pygmies, and what is the connection with my research? For more than 99 percent of its history, the human species has survived by hunting and gathering, and the pygmies are one of the last remaining peoples to live this way. I wanted to study them to understand certain aspects of human evolution in that very long period, under conditions which have now almost disappeared. Even in the 1960s, few peoples of this kind were left to study. I also wanted to analyze the then unknown relationship between the evolution of the pygmies and that of other African peoples, including the reasons for the difference in height.

To analyze the pygmies' genetic composition, I needed to obtain some blood samples. It is not always easy to persuade the average European to accept a needle prick and see his blood flow from a vein, so I had no idea how these people would react.

Our base for the first expedition was a laboratory set up by the Paris Museum of Natural History near the southwest border of the Central African Republic. It comprised a small group of stone buildings in a clearing in the heart of the forest, in an area inhabited by several pygmy groups. We used Jeeps and Land Rovers to travel around.

My first attempt was an out-and-out failure. I had arranged a meeting with a pygmy group through the head of a group of workers on a coffee plantation. The farm was owned by a provincial French aristocrat who had promised to help. I presented myself on the appointed day, with my companions and equipment, only to find that all the pygmies had disappeared into the forest. Word had spread that I was a *likundu* (a demon or bad witch). They left their village idiot behind for my tests—I never learned whether this was done in scorn or to see what I would do to him.

I worried that my bad reputation, acquired without warning or explanation, would spread and follow me, so I decided to look for other pygmies as far away as possible. That first year we were not equipped to stay out overnight, so the area that was seven hours away by Jeep was the farthest we could travel. We left very early (at 1 A.M.) and returned late at night. I remembered a film on the ivory trade, which I had seen as a child, mentioning that the pygmies adore salt; so I took a great deal of salt with me and this time avoided intermediaries.

The approach worked wonderfully. Taking blood samples from the pygmies became extremely easy, much easier than from any other people I worked with before or since. We took gifts of salt, soap, and tobacco, and medicines to treat some of their illnesses, but I believe that the respect we always showed also contributed to our good relations. Pygmies are very good at distinguishing friends from enemies and recognizing the real intentions of others.

In the last twenty years, I have been to Africa ten times and taken blood from more than fifteen hundred pygmies in thirty different places, not just in the Central African Republic but also in Cameroon, Zaire, and other African countries. I have returned to some of these places several times and have always been greeted enthusiastically.

The Peoples of the Forest

Paleolithic society must have been organized very much as the pygmies are today. They are nomads or seminomads. A tribe numbers five hundred, one thousand, or two thousand people, sometimes more, always living in groups of about thirty, on average (although it can vary from ten to fifty, including the women and children), who hunt together. Every so often the groups—or whole tribe—meet to celebrate great dances and collective rites. Dancing and singing are the main social activities.

Because a pygmy house doesn't take long to build, frequently moving camp and setting up a new one a few days's journey away is easy. The composition of the village is fluid. The few component families are usually, but not always, related on the male side. With every move, some go their own way and others arrive, making each new settlement different from the one before. Hunting territory is split among the groups and passed down from parents to children. The right to hunt in a different area can be acquired through marriage. About 30 to 40 percent of the pygmy diet is game of some sort, usually antelope or gazelle. Monkey is

1.1 Loincloths made of beaten and painted tree bark. Work of the Epulu pygmies from the Ituri forest in Zaire.

considered a delicacy, particularly our cousins the chimpanzee and gorilla, which inhabit the pygmy areas. The men hunt, and the women gather the rest of the food (fruit and all sorts of vegetables).

They go barefoot and, until a few years ago, were almost totally naked, their only attire being a loincloth, generally made from the bark of a tree. They do not weave and whenever possible they obtain cotton cloth or ragged shirts and trousers from local cultivators. When I began my research in the mid-1960s, they still made their loincloth out of bark that had been beaten to soften it. The pygmies in northeast Zaire sometimes decorate them with beautiful patterns; these may fetch hundreds and even thousands of dollars on European and American markets.

The pygmies are extraordinarily well adapted to the forest and are experts on every living thing. They use herbs and roots to make medicines generally unknown to Western doctors. They dip their arrowheads in a deadly poison, using extracts from three of four different plants, and have even developed the antidotes for these poisons. What they know most about, however, is ethology—animal behavior—which is fundamental to a hunter. They are the only humans able to survive on their own in the forest. Years ago I saw a very good and scientifically accurate film in which a pygmy showed a child how chimpanzees catch insects by using a stick to open the tunnels in their nests in tree bark. The termites are disturbed by the unusual condition and swarm over the stick in wild agitation, upon which the chimpanzee quickly withdraws it and gleefully eats his meal. The discovery that chimpanzees manipulate instruments—of which the stick used to catch termites is the most important—caused a great stir some years ago. It was made independently by the British ethologist Jane Goodall, who after months of work was finally accepted by a group of chimpanzees and spent years studying their habits and customs. The pygmies knew all about the chimpanzees centuries ago, even if they prefer their termites cooked.

Pygmy Life

Knowing the pygmies has been an extraordinary experience for me; they are the most peace-loving people I have ever met. Polite, very dignified, and even humorous, they hate and avoid violence. When they disagree, they argue and may hit each other—even husbands and wives —but they rarely use weapons. Murder is extremely rare. If two people disagree, they avoid each other, stop speaking for a while, and build their huts so the entrances do not meet. If the dispute is more serious, one of them will leave the village to join another group.

A fixed rule in pygmy ethics (possible only in large, sparsely populated areas!) is that people who quarrel badly must separate. Their companions, tired of hearing voices raised in anger, endeavor to silence them. If the arguers persist, they are sent outside the village. Pygmies hate those who "make a noise" or "disturb the peace," to use their own words.

Men and women are equal in pygmy society, and everyone discusses matters of interest around the fire. The most serious punishment that can be inflicted is expulsion from the village, which amounts to a virtual death sentence. Life in the forest is fine in a group, but it is nearly impossible to survive alone. Sometimes, of course, another group will accept the exile.

One remarkable feature is the parents' exceptional love for their children. And although parents raise their own children, all the adults treat the children as if they were their own. A child who is orphaned is immediately adopted without distinction by a family of aunts or uncles. Colin Turnbull, the first anthropologist to live at length with these people, and an excellent writer, notes that the children call all adults of their parents' generation "mother" or "father," all those of the generation before "grandmother" or "grandfather," and all those of their own generation "sister" or "brother." The old and infirm are protected carefully, as long as they do not jeopardize the group's lives. I remember one pygmy being called to a farming settlement to kill a mad gorilla —the locals often rely on the skill and bravery of the pygmies. Having mortally wounded the gorilla, the man was bitten very severely in the lumbar area and paralyzed from the waist down, but he was kept in the tribe. (In the forest, inability to walk means certain death.) The pygmies take care of the injured, and I have seen blind and extremely sick people kept within the group and cared for by their relatives.

I am often asked how long pygmies live. Their life expectancy at birth is about seventeen years, which sounds terrible compared with that of U.S. males (70) or females (76). Many deaths take place at very early ages, because of infectious diseases. It is rare to find individuals above 60 years of age, but the ages of individual pygmies is usually based on estimates because they have no interest in it. In spite of high mortality, they maintain (barely) their numbers. Other Africans are better off; even if health services are usually very poor, many of them have some, however limited, access to modern medicine; pygmies essentially have none.

The pygmies no longer have a language of their own. They speak the language of the peoples with whom they have come into contact over time, perhaps even centuries previously. Because they have also moved long distances, their language may be borrowed from peoples living far away. Turnbull says they appear to attach no importance to the future or the past; what counts is the present. As they say, "If it is not here and now, then what do where and when matter?"

Their god is the forest, of which they feel themselves an integral part. It is both father and mother, the being that permits them to live and that must be respected. Depending on the region, after death pygmies are either burned or laid in their hut, which is knocked down around them after a funeral rite. Their bodies are then left to dissolve into the earth while their companions move on.

Marriage is not a very complex rite. If need be, pygmies divorce. The current habit of "buying" their wives probably was adopted from local cultivators. They pay not with money—they do not have money—but by working for the future parents-in-law, perhaps doing their hunting for a year or two. Before he can marry, a man must prove he can hunt game and therefore support a family. When he takes a wife, he must also give something in exchange, to replace the contribution the woman made to her family.

The Pharaoh's Message

Pygmies have a cheerful temperament: they are chatterboxes by nature and—rightly enough—feel they lead a good life. They often stay in the settlement doing nothing. They also are keen dancers and singers; and their music has an extremely rich range and unusual timbre. Each pygmy repeatedly produces a note or a certain tune at a regular time interval. They have an incredible sense of rhythm. A French musicologist once taped a solo piece of music in which the same note was repeated at an irregular interval, and then he taped the same piece performed by another and then a third man. Superimposing the three recordings, he found he had a choir in perfect union. Pygmies use simple musical instruments such as drums, flutes, and a kind of one-string violin. Some areas boast excellent groups of musicians.

Their overwhelming passion, though, is for dance: even before they can walk, six- or seven-month-old babies will dance to music if held up by their hands. Mothers with small children dance with them on their hips or shoulders. There is always a drummer willing to tap out a beat and someone to show off with fast and difficult steps. In a letter, a pharaoh of forty-three hundred years ago exhorts a general in search of the source of the Nile to bring him a pygmy from Punt (perhaps Ethiopia or the Upper Nile), calling him "God's dancer" and giving instructions to treat and care for him so that no harm will come to him.

All these years later, the enthusiastic tone does not surprise those who know the pygmies. They normally play their music in a wide circle around the fire and will go on singing and dancing in their lively, agile style all night long.

There is a pygmy dancer by the name of Aka in an Egyptian fresco. Even today, one pygmy tribe uses the name Aka to define itself. This name, which has survived many thousands of years, quite simply means "human."

The Shortest People in the World

Since ancient times, the pygmies have been known as the world's shortest people—even Herodotus and Aristotle mention the fact. In reality they are not as small as many people believe. In the smallest tribe, the average height of the men is 4'9" and the women 4'6". Some pygmy groups are as much as 4 inches taller. Today we can still meet people every bit as small as an African pygmy, but it is only on seeing several of them together that we realize they are unusual. They should not be mistaken for dwarves, who are short as a result of certain physical disorders. Some forms of hypopituitary dwarfism lead to even shorter statures.

We don't know if the pygmies have become smaller over time or if they have always been small. If, as is possible, they have always lived in the forest, there is no hope of finding their remains, because the soil is so acidic that even bone dissolves rapidly.

Early humans, dating from two or three million years ago, were very small. In the industrialized world, average heights have increased over the last two centuries, mainly as a result of improved diet. Medieval armor would be much too small for a person today. The first systematic data on stature date back to the early nineteenth century when Napoleon had his conscripts measured for height. Our great-great-great-grandparents were much smaller than we are, our great-great-grandparents a little less so, and so on. The biggest leap has been made in this century, first in northern Europe, then in the south. Have the pygmies become taller too? Some say yes, but we cannot be sure.

1.2 The Scottish anthropologist Colin Turnbull with Makubasu, an Epulu pygmy from the Ituri forest in Zaire. Height: ca. 5'11" and 4'7", respectively.

Why Are They Small?

The pygmies of today all live in the African forest. People who live in tropical forests, where the climate is very humid, are generally very small. Examples are the peoples of southern India, Indonesia, the Philippines, and New Guinea, as well as the Maya of Central America and the inhabitants of the Brazilian tropics. The pygmies, though, are the smallest of all.

The climate of the equatorial forest is unusual—similar to that in a greenhouse, it is not very hot, but humidity stays around 100 percent. Although it is not very warm, the humidity is so high it prevents the body from shedding internal heat. Under normal hot conditions with lower humidity, one sweats intensely, which helps cool us down since perspiration chills as it evaporates. The same mechanism is used to keep food cold: A special fluid evaporates in a closed container, absorbing heat from the inside of a refrigerator. It then returns to its liquid state, only to be evaporated once more, in a continual cycle.

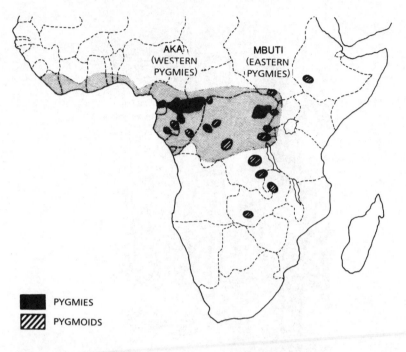

PYGMIES
PYGMOIDS

1.3 Equatorial forests and surviving pygmy groups in Africa.

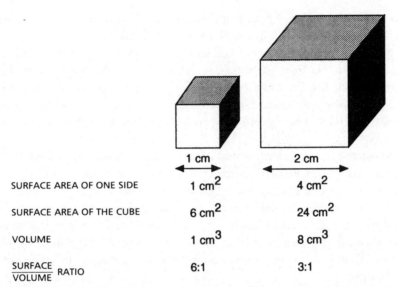

	1 cm	2 cm
SURFACE AREA OF ONE SIDE	1 cm^2	4 cm^2
SURFACE AREA OF THE CUBE	6 cm^2	24 cm^2
VOLUME	1 cm^3	8 cm^3
$\frac{\text{SURFACE}}{\text{VOLUME}}$ RATIO	6:1	3:1

1.4 When body volume increases, the ratio of surface to volume diminishes and heat generated by the body is lost more slowly.

But near 100 percent humidity, the body's normal ability to cool down is hampered or stopped completely. If perspiration does not evaporate, it does not cool the body. In such critical conditions, we risk overheating, with the body temperature rising above the normal 98.6 degrees Fahrenheit or passing the survival limit of 107 to 108 degrees. Heatstroke may be fatal and we must protect ourselves in some other way.

Pygmies sweat a lot, but not enough—it is their small stature that protects them, in two ways. First, the surface area of a small body is greater in relation to its volume. It is a mathematical fact: if cube A in Fig. 1.4 is 1 centimeter along each side and cube B is 2 centimeters, then A's surface area is one-quarter that of B, but its volume is eight times smaller. Heat is produced in the mass of the body, particularly in the liver and muscles, and is lost through the surface; if the latter is larger relative to body mass because a person is small, heat loss is easier and cooling more efficient. In a warm and humid environment, it is best to be small.

Second, small stature is an advantage if a person must use a lot of energy, since less energy is needed to move the small person's body weight. Athletes can produce large amounts of heat—marathon runners, for example, have to sustain intense muscular effort over a long period; they are usually relatively short, although it may seem better to be tall because of the longer stride. Pygmies use less effort to move than a larger people because the weight they have to shift is smaller.

Similarly, ponies are more efficient than large horses in terms of the energy produced from the food consumed. You need a big horse to shift a large load; but for traveling or drawing a small cart, a small horse or even a donkey will do. In the nineteenth century, U.S. transport companies used pack horses, but express mail delivery was performed by ponies.

Small stature, therefore, would appear to be an adaptation to life in the forest, which may have taken many years, at least three to five thousand. If, as some say, the tropical forest did not exist five thousand years ago, it may be that the pygmies adapted over this period. It is, however, possible, and I believe probable, that the forest already existed and that the pygmies have been there much longer. There are no fossils as proof, and we do not know enough about the climate in other times to be sure.

Although their bodies are small, the pygmies' heads are the same size as ours. The torso is muscular, and the arms and legs are slim and tapered. Their legs are rather short, but the overall effect is graceful. They are athletic: the men can climb trees one hundred or more feet with amazing agility.

They have elongated eyes and the widest nose in the world, also an adaptation to the forest. Small nostrils are useful when it is cold, since air has time to heat up before it reaches the lungs. In the forest, however, the air is warm and humid; there is no need to alter temperature and humidity through the filter of the nose, so wide nostrils are better.

Pygmies and Cultivators

Nowadays many pygmy groups no longer spend the whole year in the forest. Four or even six months of the year, in the dry season, they build their huts outside a settled village of cultivators who use them as plantation labor and treat them as land servants.

In reality, many cultivators consider the pygmies less than human. They have created a system of hereditary service: a pygmy always works for the same master and subsequently for his heirs, and the pygmy's children go on working for the same family. This form of serfdom exists mainly in the intentions of the cultivators, and occurs only when the pygmies consent to it. If not treated reasonably, they disappear back into the forest, where they cannot be found. It is in the cultivator's interests to take care of their pygmies and establish an acceptable relationship. Their attitude generally is arrogant, but occasionally is affectionate as well.

The economic arrangement between the two groups usually works to the disadvantage of the pygmies, who do not use money and have no conception of its value. The cultivators generally don't know how to hunt or raise animals, except chickens, a few goats, and more rarely pigs, so they obtain meat and other products from the forest, mainly from the pygmies. In exchange the pygmies receive, apart from food, iron goods, such as spearheads and knives, and terra cotta pots (which are increasingly being replaced by saucepans made in China). They do not know how to work iron themselves, because their nomadic lifestyle does not allow them to carry heavy items such as anvils and forges.

The pygmies are considered the poorest of the poor, the occupants of the bottom rung of the economic ladder. As far as possible, the cultivators make sure they remain unaware of the value of money, for fear they will become too expensive. In the opportune season, at least, the pygmies do a lot of field work for them. They also, being light and agile, build house roofs. Apart from tools, the pygmies are paid with alcohol, tobacco, bananas, and cassava (manioc), which until recently were produced only by the farmers.

Now some of the pygmies are beginning to grow cassava, because it requires very little supervision—just plant a sprig and return after two years to gather the roots. Even the leaves make a tasty soup. A native of South America, since its introduction two hundred years ago it has gradually replaced the majority of previous crops. It is not only easy to grow but (like wheat, potatoes, and rice) does not clash with other flavors.

In any case, the pygmies don't farm willingly—they farm only when the forest is destroyed. Then, forced to change lifestyle, they may take up pot making and fishing, as well as farmwork. They survive as best they can, but continue to hunt whenever possible.

Farm life has never been considered much fun. In the Bible, when Adam and Eve taste the apple, God drives them from Eden to make them "work the earth of which they were fashioned," saying, "in the sweat of thy face shalt thou eat bread."

Hunters generally work less than cultivators. This is true also of the few hunters left in the savanna, with its vast population of herbivorous animals. These grasslands have been almost completely swept away by developed economies, but they once must have been a fine sight with their rich variety of highly visible game. In the forest, it is easier for animals to hide.

The pygmy women's work is perhaps more monotonous, although they seem content with their lot. The cultivators' wives, who do the heaviest labor in the fields, work decidedly harder than the pygmy women.

The forest may look gloomy to us but pygmies feel entirely at home and safe there. It is a place where little that is untoward can happen to them, where danger is limited and life very pleasant. The same applies to all the modern hunter-gatherer groups whose culture and history we know. They are (or were) extremely well adapted to their environment, but when it changed, they inevitably had to change too, or disappear.

The Hunter-Gatherers of Modern Times

Before agriculture took over, there were a few areas in which it was very easy to hunt, gather, and especially fish. For example, in the great river estuaries in the North Pacific, in North America, the salmon were so plentiful in season that people could catch them with their hands. Once smoked, they lasted a long time. Another prehistoric people who became numerous were the Japanese, who lived by hunting and acorn gathering but also by fishing. In particularly favorable environments, these fishing societies reached population levels hundreds or even a thousand times denser than the great majority of hunter-gatherer communities.

Before agriculture, the population of the earth was unlikely to have exceeded five or ten million people. It has been calculated that in England there were probably five, maybe ten, thousand inhabitants (almost ten thousand times less, that is, than today). The adoption of agriculture led to a demographic explosion. World population has increased one thousandfold in the last ten thousand years. Even in traditional African societies, those tribes who became cultivators several thousands of years ago (in Nigeria, for example) now number many millions of people. The tribes that stuck to the oldest style of living, hunting and gathering are still very small in number even now.

Before white colonization, farming societies often lived side by side with hunter-gatherers who perhaps practiced a little cultivation. The North American Indians, for example, were mainly hunter-gatherers. In

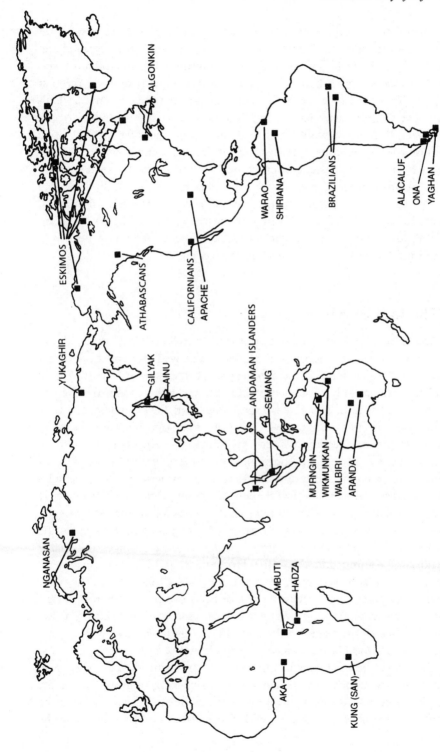

1.5 Ten thousand years ago, all the earth's inhabitants were hunter-gatherers; these numbered no more than a few million in all. The map shows the most important of those that still survive or have only recently abandoned hunting and gathering or died out.

the plains, they lived by hunting buffalo, which became much easier after the Spanish brought horses to the continent. Some of these escaped and returned to the wild, where they were captured by the Indians, who began to use them to hunt. On their arrival in North America, the first whites discovered that the natives had already developed some agriculture. The pilgrims who landed in Massachusetts at the beginning of the seventeenth century would not have survived the first winter alone because their provisions were insufficient. They procured corn and other foods from the Indians and later learned how to grow the local crops.

In South America, too, there were hunter-gatherers, but Central America and the Andes, along with a slice of the great plains and forests, were already highly organized in agricultural terms. At the southern tip of the Andes, especially in the Tierra del Fuego, land was poor and inhospitable, and agriculture had not arrived.

The Last Survivors

About five thousand human populations remain on the earth, judging from the number of languages in existence; this is an imperfect but useful measure, if only because the languages are the sole aspect of culture studied systematically enough. Very few hunter-gatherer societies are left —about thirty until some years ago, now even fewer. Today the only numerically important groups (with more than one hundred thousand members) are the central African pygmies, the Khoisan of southern Africa, and the Australian Aborigines; only the pygmies, and not all of them, live predominantly by hunting and gathering. Other smaller groups exist, scattered about. Of these we know, if not the lifestyle they once pursued, at least what remains of it after being disturbed by us.

There are two populations in South Africa generally known as the Bushmen and Hottentots. Anthropologists have coined the name "Khoisan," joining *Khoi-khoi* (the Hottentots) to *San* (the Bushmen). Today they inhabit very arid areas, but previously they lived in the more pleasant savanna, from where they have been driven. Apart from differences of environment and ecology, the Khoisan live, or lived, in a way fairly similar to the pygmies. Few are hunter-gatherers today. Many are farm laborers; others are soldiers or somehow live in the city.

They have preserved their language, which is undoubtedly extremely old and very unusual: it contains unique clicking sounds, which are highly varied and difficult to repeat. Various South African tribes, such as the Xhosa, have mixed a lot with the Khoisan, a fact proved by the appearance of three new clicks in their language and of Khoisan genes in their DNA. The Khoisan also have characteristic, almost oriental, features. Nelson Mandela, Africa's most charismatic politician, is of Xhosa origin, and his face shows clear signs of Khoisan influence. The word *Xhosa* really should be written !*Xhosa*, the exclamation mark indicating a special click in phonetic conventions.

The Australian Aborigines are dispersed or live together in reserves, where they have a more sedentary life than before, except for a group in the north. To these, the government has assigned a territory in which they can continue to live as in the past. There are about 170,000 Aborigines, of whom 47,000 still preserve some knowledge of their original language. Only relatively few live in ways similar to the traditional ones. In the past, they lived in groups of twenty-five to thirty people, and the tribes, when they existed (the majority have been destroyed, broken up, amalgamated, or put into reserves), numbered four hundred to five hundred people on average. Each surviving Australian tribe has a different, ancient language (not borrowed from neighbors, as is common with pygmies). When James Cook arrived at the end of the eighteenth century, agriculture was still unknown.

These average dimensions for groups of hunter-gatherers are the same as those found on archaeological sites that precede the discovery of agriculture: the settlements and campfires along with the amount of animal bones found indicate numbers equivalent to the groups of pygmy hunters and Australian Aborigines. Naturally, the hunting band, the approximate equivalent in size to a modern pygmy and ancient settlement, varies in number because it is dynamic and mobile; it may tighten to form smaller nuclei or widen to a larger group according to the demands of hunting and survival. The tribe is a social unit superior to the band. It is more stable and tends to be endogamous, meaning that most marriages (80 to 90 percent) are contracted within it. It has also been found that five hundred is the minimum threshold for easily avoiding excessive intermarriage, with its damaging effects on offspring.

The pygmies say it is best to "marry afar," because the man acquires hunting rights in his wife's territories. This way he broadens his hunting ground and prospects for survival. This is a highly valid economic motive, but the custom is positive for an additional extremely important, if unconscious, reason: it reduces the chances of marriage between close blood relations and enriches the local genetic pool. Primitive peoples living in small groups run serious risks of inbreeding, leading to lower vitality and infertility in the children. Custom almost always finds a way of avoiding this danger, however. A long-term genealogy has been reconstructed for the Thule Eskimos, a tiny, rather isolated group inhabiting the far northern coast of western Greenland. An extraordinary series of genealogical acrobatics has ensured that although marriage inevitably takes place between relations, they are always distant ones.

An Example of Exhaustion of Genetic Variation

Any group, even the most primitive, is genetically highly varied. Customs preventing inbreeding help keep even the smallest human population rich in genetic variations. The only example I know of a human population in which genetic variety was apparently partially exhausted is that of the tribes of the Indian-controlled Andaman Islands, near Burma.

There were at least four Andaman tribes, who spoke different, but related languages. In the nineteenth century, these tribes were still made up of a reasonable number of people, and at least one still numbered five or six thousand. Contact with whites, and the British in particular, has virtually destroyed them. Illness, alcohol, and the will of the colonials all played their part; the British governor of the time mentions in his diary that he received instructions to destroy them with alcohol and opium. He succeeded completely with one group. The others reacted violently.

Ancient travelers give differing accounts of the Andaman tribes' behavior toward visitors. Marco Polo said they were fearsome, but, because he also says they had dogs' heads, I doubt he had been to the islands himself. Giovanni dal Pian del Carpine, another Venetian who preceded him, states that they were most well mannered. The truth may be that the different tribes had varying reactions to outsiders. In the mid-1800s, when a British ship visited Little Andaman Island and sent a launch ashore, the native Onge captured some of the sailors, cut off their arms and legs, and then burned the still-living trunks on the beach before the eyes of those who had managed to escape back to the ship. The British reaped terrible revenge a year later. A group of soldiers went ashore and waited until the natives came onto the beach; they then opened fire and killed about seventy before returning to their vessel.

There was no further contact between the Onge and Westerners until the twentieth century. In 1951 an Italian anthropologist, Lidio Cipriani, went to the island for two years and conducted some excellent research. He managed to gain the people's trust and became the first to provide accurate information. He explained, for example, why the Onge mutilated the British sailors and then burned the trunks. According to their religion, if they had not done so the spirits of the dead could have returned to haunt them.

These people were extremely primitive. They had even lost the ability to create fire. They knew how to protect it, but could not produce it. The pygmies, however, know how to make fire, although they find it easier to keep it lit. The women carry embers with them through the forest.

The numbers of Andaman natives had already declined considerably, even without the British. Today the Little Andaman Onge number no more than ninety-eight or ninety-nine people, too few to avoid close inbreeding. The result is that most couples have no children, or have, at most, only one or two. They are careful to ensure the tribe's survival, so if a girl has no children from a first marriage she is taken and married to another, and then another if necessary.

Another group of Andaman natives, on tiny Sentinel Island, indicated in January 1991 that they were willing to make contact with the Indian government for the first time. I suspect that these are the last native people never to have had contact with the rest of the world.

Very Different Rules of Conduct

All of today's hunter-gatherers have certain customs in common. They live in small groups, have no social or political hierarchy, are normally without leaders, and base their social contacts on reciprocal respect.

Their rules of conduct are usually highly developed. The people at the bottom of the economic ladder are in no way primitive, ethically speaking; they merely see things differently from us. When the Dutch settled at Cape Town, their herds spread into the native hunting territories, causing incidents with the Khoisan. Exactly what occurred is not clear, because there were no anthropologists at the time, but that the natives were irresistibly tempted by the Dutch cattle is not surprising. The Boer farmers started to shoot the Khoisan on sight and literally exterminated them in wide areas. They now live only in Namibia and Botswana, in deserts, in poor, unwanted savannas, or in city ghettos. The Hottentots in the Boer areas have survived better, probably because they already raised cattle and did not hunt other people's cows.

Hunter-gatherers have a different view of possession, because private ownership is rare and not very important. A few rights, such as those regarding hunting territories, are upheld. A pygmy caught hunting on someone else's territory is fined, not in money, since it does not exist, but in kind. The Pygmies have no respect for the cultivators' property and—if they can get away with it—steal the bananas and cassava grown using pygmy labor in the fields and so frugally shared by the farmers. In reality, the enclosures of today are their ancient hunting grounds, forest areas cleared without permission or recompense. The pygmies know that the farmers exploit them and consider them animals, but they cannot afford to break off the relationship. They take revenge by stealing food whenever they can. Turnbull says the pygmies have two paths into the forest—a hidden one they use themselves, and a wider more direct one for the farmers who come looking for them, which the pygmies use as a toilet hiding their excrement under leaves. I was struck by the episode of a pygmy, mortally wounded by a cultivator who caught him stealing bananas. The cultivator was arrested and condemned, but, before dying, the pygmy asked his pardon and forgave him, saying it was his own fault since he should not have stolen the bananas. As far as I am aware, he had not been influenced by Christianity; neither Catholic nor Protestant missionaries have had much access to the pygmies. The big, generally brick missions are in the more important cities, far away from pygmy territory. At the most, mass is held once a week in one of the bigger outlying villages.

There are exceptions. African pygmies are commonly affected by a disease called yaws, caused by a spirochete very similar to that responsible for syphilis. Unlike syphilis, it is not a venereal disease, but it is transmitted through skin contact. Frequently fatal to children, in adults it causes mutilations almost as serious as leprosy. It can be cured with a single injection of penicillin. A small Milanese nun who worked for years at the Nduye mission in Ituri, the main pygmy region in Zaire, used to walk for days through the forest to give penicillin shots in the farthest pygmy camps.

Societies Without a Future

These few remaining primitive societies have hardly any future. Reduced to small groups, often only a few hundred or thousand, they are undergoing the destruction, by industrial and monetary economies, of their habitat and traditional way of life.

In a northern Australia park, the Aborigines still carry on according to their customs. Some manage to improve their condition by doing artwork. Nearly all the others, however, have abandoned the traditional lifestyle and live mostly unemployed in the barrack villages.

The Eskimos, who number twenty thousand in Canada and fewer still in Alaska, survive through social subsidies. Some work on the radar stations, or live off artwork. They hunt, but are unlikely to use kayak and harpoon. If they have the money, they buy a rifle and outboard motor. They have not lived in igloos for thirty years. Their homes are prefabricated barracks supplied by the government.

With the help of art dealers and the Canadian government, they have developed sculpture using a beautiful local stone. The Canadian artist John Houston lived with them for nine years to teach them how to make lithographic prints, without influencing their traditional tastes. Few have a genuinely original style, of course—most simply produce salable statuettes or so-called airport art. It is striking to think, however, that over 70 percent of Canadian Eskimos have at some time tried to earn a living through this kind of work. In some villages today, everyone is a practiced recognized artist.

With a few minor exceptions, the Lapps, too, are no longer hunters and fishers. Some hunt tame female reindeer, which they use to bait the males, but their lifestyle is not exactly primitive. A Swedish psychiatrist who is a friend of mine recently went to a northern area to help a group of schizophrenics. On attempting to visit some Lapp friends, he was told they had gone fishing—in a helicopter!

Pygmies still hunt with the bow and arrow, and in some areas with the crossbow, but mostly with the net. Theirs is the future of the forest, destroyed slowly in the past and rapidly now. In Africa the process began three thousand years ago, when the Cameroon Bantus spread south and east, and has continued slowly ever since.

The Amazonian forests of Brazil are similar to those of Africa. In fact, they are a continuation of each other beyond the Atlantic Ocean, and they were actually linked before the separation of the two continents one hundred million years ago. In recent years, the Amazonian forest has been destroyed at a dizzying rate, with disastrous consequences for the hundred or so Indian tribes living there.

Respect for indigenous people who still live in traditional ways is unfortunately rare or totally lacking. Even the names we give them betray our lack of respect. For ease of understanding I used the names that everybody is familiar with. A student of anthropology in Oregon, who is a Lapp, was telling me, almost crying, that Lapps means, to neighbors, "good for nothing," because Lapps do not cultivate the land. Lapps call themselves Saame. Similarly, Eskimos call themselves Inuit, Bushmen San, Hottentots Khoi, and these are the names we should use. Pygmies call themselves in many different ways, but several tribes are called Aka, and others Twa. Probably these names have the same origin: Aka-Akwa-Kwa-Twa.

Attempts at finding new, more comprehensive names have often misfired. A suggestion to use the name Khoisan for Bushmen and Hottentots makes sense linguistically because their languages belong to the same family, but some Khoisan people have objected to it.

We may try to show respect by using more proper names, but the real measure of respect is another one. Our civilization will be judged eventually, among other things, by the way we understand and help indigenous people. So far, our score is extremely low.

Why Investigate These Strange Peoples?

We know that about two million years ago there lived the species *Homo habilis*, whom we recognize as human, albeit very primitive. These people descended from ancestors shared by both the humans and chimpanzees of today, who lived about five million years ago. We know of other ancestors who lived even earlier. But *Homo habilis* so far remains the oldest find whose features qualify this species as part of humankind: these people no longer used their hands to walk (their ancestors already did at least two million years before), but to make tools.

Since then the human species has slowly evolved until it has become "anatomically modern human," the only category of human alive today. Modern humans appeared only some hundred thousand years ago, but until a few thousand years ago—quite recently, that is, in terms of human evolution—all *Homo habilis*'s descendants lived much as their ancestors had, in small, seminomadic groups, hunting, gathering, and fishing to eat, and using stone tools (and probably wood, which have very rarely come down to us). Change was slow. In the last few thousand years, however, change has been enormous: crop and livestock farming have been invented, leading to an explosion in population density. A thousand times more humans live on the earth now than ten thousand years ago. Lifestyles have changed enormously too, except in small enclaves with extreme ecological conditions, such as the tropical forests and arctic tundra, where the raising of crops and animals is difficult, impossible, or in any case economically unviable.

Studying the pygmies, Khoisan, Eskimos, and Aborigines helps us understand how our ancestors lived, although obviously the lives of these peoples have changed. For at least two thousand years, the pygmies have had contact with cultivators, and barter their game for iron tools, which are better than stone ones. When the whites arrived in Australia, the Aborigines still used stone tools, similar to the oldest ones, and to some extent do so today. Visiting the New Guinea plateau in 1967, I found stone tools still being used, although their design was more recent.

House-building techniques and social life of modern hunter-gatherers, as well as numbers, probably have not changed much, along with the habit of sharing food, the absence of hierarchy and established laws, and probably also customs linked to fertility and birth.

Ethnography, the study of peoples living in socioeconomic and technological worlds other than the industrial one, is extremely useful in

understanding the making and use of tools found on archaeological sites. We admire the cleverness of these people. The hunter-gatherers are the most interesting, since their lifestyle is the closest to that of ancient humans. The geneticist is particularly interested in their population structure, especially how people find mates, which helps us understand the degree of variation among nations, tribes, and villages. The differences between their structure and a modern society's provide precious information on the evolution of the human species and the differences between ourselves and our recent ancestors, who also fall into the category of modern humans.

Portraits from the Past

In the early seventeenth century, James Ussher, archbishop of Armagh in Ireland, determined the date of the creation of the world at 4004 B.C., based on a scrupulous analysis of the Bible.

Until the last century, common belief put the span of humans at only a few thousand years. It was thought that God made humans not long before the start of history, which is set at about five thousand years ago, at the time of the first Sumerian and Egyptian documents.

We now think that agriculture was invented ten thousand years ago, the species *Homo sapiens* is about two to three hundred thousand years old, while life has existed on earth for three or three and a half billion years. A long series of finds that were initially met with skepticism has changed our ideas on the birth of humans and the planet. In his time, Archbishop Ussher's convictions were widely accepted, and some groups of Christian fundamentalists still fiercely support a literal interpretation of the Bible.

A Controversial Discovery

Scientific analysis of fossilized human remains began only in the last century. Paleontology, or the study of fossils and ancient life forms, is older. Herodotus cites shells found in the Egyptian hills to prove that Egypt was once undersea.

Certain fossilized human remains had already been found at the start of the 1800s, but not until the discovery of the first Neandertal, in 1856, did the world realize that there may have been humans different from us in ancient times. Like so many, this discovery was made by chance. During quarrying, some unusual bones came to light and were handed to the quarry owner, who entrusted them to the care of the local schoolmaster. This occurred near Düsseldorf, in the Neander River valley; *thal* is German for valley, but the *h* is dropped except in special usage. Realizing that the find was significant, the schoolmaster passed the bones to Hermann Schaaffhausen, a professor of anatomy, who found the bones striking enough to inform the scientific world. Many maintained that they were not unusual; others suggested strange explanations, including that they were Cossack bones.

The bones resembled normal human ones, but were thick and heavy and much stronger. The cranium was long, low, and flat, and the brow ridges were large, very prominent, and arched. It was easy to make mistakes (science as a whole, and paleoanthropology is no exception, has a history full of errors) because the find was indeed out of the ordinary and completely in contrast with religious beliefs that humans had been created, thereby precluding the possibility of any subsequent evolution. When the first dinosaur bones were discovered, again in the nineteenth century, it was said that they could not belong to an extinct species because God would not have made a mistake in inventing an animal and then rectified it by making the animal die out. Germany's greatest pathologist, Rudolf Virchow, maintained that the bone shape of Neandertals was the result of an illness, and his prestige was such that many scientists who had previously expressed opinions similar to those we hold today actually changed their mind.

It was some time before the idea that the bones belonged to primitive humans became accepted. The person most influential in soundly promoting the idea that humans and apes could have evolved from a common ancestor was the naturalist Thomas Huxley, a friend and supporter of Charles Darwin, the world's most important evolutionist.

Darwin himself was more cautious than his friend when speaking of a common origin for humans and apes, probably fearing that a massive backlash from the Christian churches would crush his theory before it could become widely known. Huxley crossed swords openly with bishops and politicians, and often prevailed by dint of his great oratory talent and sheer intelligence. When Bishop Samuel Wilberforce derided him, saying that both he and Darwin obviously descended from the apes, Huxley replied that he much preferred being related to the apes than to men who reasoned like the bishop.

The Discoveries Multiply

In the decades that followed, many other Neandertals similar to the first came to light, proving that humans different from us really did exist. Some finds predated the 1856 one but had not been publicized.

In 1868, excavations for a railway line in the Vezere River valley, in the Perigord region of southwestern France, brought to light other human remains. These are very old, too, but are decidedly closer to us in appearance. As usual their name derives from the place where they were found, Cro-Magnon, which is not far from the places in France where Neandertals had been found. Seventy years later, the Lascaux caves were discovered in the same region, their walls covered by extraordinary frescoes depicting prehistoric animals and hunt scenes, work of the so-called Cro-Magnon race. The Lascaux discovery was made by a group of young people who followed their dog through a hole that led them into the caves.

Cro-Magnon resembles us closely and is what we call a modern human —if we met one in the street, he or she would not seem strange, whereas a Neandertal would probably strike us as odd.

Other, even older humans than Cro-Magnon and Neandertal have since been found. In the late 1800s, a Dutchman found human remains in and around Java that almost certainly predate Neandertals. A body of information that could no longer be ignored or buried by prejudice was gradually being formed: human paleontology, from being considered the product of a fevered imagination, became a recognized discipline.

Finds are often serendipitous because we have no system to show us where to look. We know that some areas—East Africa, for example—are better than others. Luckily, discoveries are publicized and usually draw great attention.

Unfortunately, much material has been lost because it has been used for other purposes. In China, for example, fossilized bones are considered a beneficial cure. Called "dragon bones," they belong to prehistoric dragons, according to the Chinese. They are sold over the counter at the pharmacy, ground down to a very fine powder. I have bought some myself. This custom has probably destroyed more than one precious find, but it is a traditional medicine much prized by the Chinese.

What Is a Fossil?

There is no precise definition of a fossil. As the word suggests, fossils are objects found underground. Most are petrified, but even some that are not can be considered fossils. The process of petrification is the slow substitution of certain chemicals by others, which are much harder, leaving the original form unimpaired. Saline solution seeps into the organic remains of a living creature and deposits salt, calcium, or silicate crystals, which replace the organic parts until the organism becomes like stone. In the end, all or most of the original material has been destroyed. The organism is chemically transformed but, incredibly, its shape remains the same. There are insect wings and plant leaves from fifty to one hundred million years ago whose structure has come down to us perfectly intact. We can see the cells through a microscope. We even find the deoxyribonucleic acid (DNA), the genetic substance that contains the information required to form a living creature.

The Missing Link

The most important discoveries of the twentieth century have been made in Africa. In South Africa, in 1924, Raymond Dart, a Johannesburg anatomist, found the first australopithecine, whom we now consider the predecessor of humans. This being is not recognizable as one of us but forms a link between ourselves and the ancestor we share with the apes. *Australopithecine* means "ape from the Southern Hemisphere" (from the Latin *australis* for south and the Greek *pithekos* for ape). These creatures are very apelike but also have some decidedly human features.

The last ancestors common to humans and the nearest apes, chimpanzees, must have lived some five to six million years ago, while australopithecines lived from about four until one million years ago. The most famous australopithecine of all is the female Lucy (from the Beatles' song "Lucy in the Sky with Diamonds," which happened to be playing in the anthropologists' camp at the time of the discovery). Lucy was found in 1974 in East Africa, on an archaeological site in the Ethiopian desert between Addis Ababa and Djibouti, not far from the sea. The exact place, Hadar, is inhabited by the Afar people. For this reason, Lucy's group is known as *Australopithecus afarensis*. There is a good reason for finding similar material such a great distance apart. The sites, in fact, lie on the same geological formation, the huge, high-sided Rift Valley, created by gigantic seismic and volcanic activity. This valley stretches from South Africa to Ethiopia.

Rift Valley research was initiated by the archaeologist Louis Leakey, who worked mainly in Tanzania and then farther north. He dug for nearly thirty years before making a series of discoveries fundamental to knowledge of our past. Others have continued excavating at the northern tip of the valley, in Kenya and Ethiopia, and have found very ancient human remains and australopithecines similar to those found in South Africa. The Rift Valley has a particularly useful feature—it is highly volcanic. Consequently, very fine dust has covered the fossils, as hap-

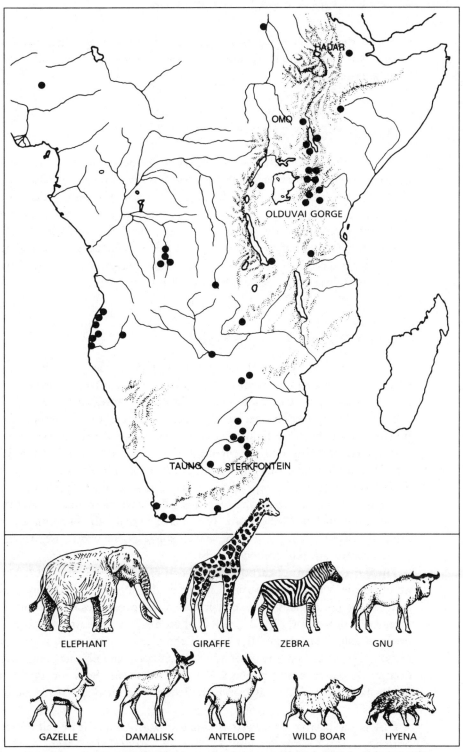

2.1 Sites of major australopithecine and early human (*Homo habilis*) fossil finds in Africa, and the animals on which they most likely fed.

pened to those at Pompeii and Herculaneum, preserving them extraordinarily well. Even the footprints of a couple, probably a man and child, have been preserved.

Dating Finds

Initial dating is done through stratigraphy. In-depth study of a region's geology reveals the order in which rock and soil layers were formed. Great care must be taken when digging to re-create the right order. The first step is to try to date the strata above and below the archaeological find—the more strata, the greater the precision. In the area where Lucy was found, the dating of numerous strata has been possible, allowing considerable precision in guessing her age. Lucy's discoverer, Don Johanson, estimates her age to be about 3.2 million years.

When Physics Lends a Hand

Stratigraphy is useful in broad terms, but particularly so for dating very old remains. Physics, biology, and more recently chemistry too, provide other methods useful when dealing with more recent periods.

The oldest and best-known dating system is the carbon 14 method. To understand this method requires a little knowledge of matter. We know that matter is composed of chemical elements, and living matter of mainly four elements: carbon, oxygen, hydrogen, and nitrogen. Each of these has several forms, depending on its atomic weight, represented by the sum of its neutrons and protons, the two elementary particles forming the nucleus of the atom. These have virtually the same mass but different electric charges. As the name suggests, neutrons are neutral while protons are positively charged.

Three major forms of carbon exist in nature: C12, C13, and C14. The number following C (for carbon) indicates the sum of the protons and neutrons in the nucleus. An element's chemical properties depend on its number of protons. C12, C13, and C14 atoms have the same number of protons—six. This means they behave chemically in the same way, like carbon. We say, therefore, that they are *isotopes*, literally meaning they occupy the same place, because they are found in the same position in the famous periodic table, which defines the chemical behavior of elements.

C14 is radioactive and exists in small quantities in nature, alongside normal, nonradioactive carbon 12 and 13. *Radioactive* means that it produces radiation: when a C14 atom emits radiation, a neutron changes into a proton. There are now seven protons. Without changing atomic weight, the atom has taken on the properties of a seven-proton atom. This is a nitrogen atom (symbol N). In practice, C14 has become N14.

The alteration takes place with absolute precision and is not influenced by temperature (an extremely important advantage). C14 is thus destroyed regularly, if very slowly, over time—so regularly that it can be used as a clock. It takes fifty-seven hundred years for the number of C14 atoms to halve. After forty thousand years, it has halved almost seven times and only the barest trace can be found. While C14 is slowly destroyed in old material, C12 and C13 remain unchanged and their amount present in a sample of matter can be measured by chemical methods. The amount of C14, even though very small, can be determined by measuring its radioactivity. The ratio of C14 to C12 and C13 decreases continuously with the passage of time in a given sample, but only organic material containing high levels of carbon, like wood and bone, can be dated in this way. If samples are older than 40,000 years the amount of residual C14 is so small that it cannot be estimated except by more expensive, and less reliable methods.

Carbon 14 is formed in small quantities in the atmosphere, a result of cosmic radiation bombarding nitrogen atoms. Together with normal carbon, it is put into circulation on earth by plants, which use carbon dioxide and water to live. The carbon 14 method assumes that the concentration of C14 in the atmosphere remains constant. This is not fully justified, but it has been possible to demonstrate the alterations in levels of atmospheric C14 at different times and rectify data. Bristlecone pine trunks preserved intact in the California and Arizona deserts for thousands of years after their death have been very useful. By counting the trees' annual growth rings, scientists have calculated the amount of radiocarbon in wood of different ages. They have pieced together the sequence of rings for a period of seven or eight thousand years, using a procedure known as dendrochronology, which takes advantage of the fact that each year the climatic conditions determine the appearance of the growth ring. The alternation of hot and cold periods creates a characteristic series of rings, to be found at least partially in all the trees living at the same time. We can now say that one tree lived, for example, between 5,110 and 5,527 years ago and another between 4,991 and 5,230.

Other chemical elements help us with longer periods: radioactive potassium very slowly becomes argon and can be used to date rock, particularly if the potassium content is high; through a long chain of elements, uranium becomes radium and, finally, lead. The uranium series can be used up to half a million years ago but is rather approximate. The potassium-argon method is useful mainly for periods between one hundred thousand and ten million or more, years ago.

The Kinship of Humans and Apes

The methods described above have helped us establish the age of fossils. How, then, have we been able to establish the link between humans and apes, suggested by Darwin and Huxley? Today we can see that apes such as chimpanzees and gorillas are our distant cousins simply by visiting a zoo. In the past, however, it was rare to see these animals—so rare that in the seventeenth century an Englishman, E. Tyson, was able to publish a book on pygmy anatomy in which the pygmy is compared to a monkey, a gorilla, and a human, when in fact the skeleton that Tyson studied belonged not to a pygmy but to a chimpanzee. The orangutan, too, is human in form, although the difference is more noticeable. Humans and orangutans separated earlier in their history.

For some decades now we have had a new genetic method for establishing the similarities between different species and dating origins. This is the "molecular clock" method, which studies various complex biological molecules and counts the differences between, say, the molecules of a human and those of a chimpanzee.

A reference point, or measure, is needed to tell us on average how many millions of years it takes for a change to come about. This measure requires a large body of data indicating when the separation occurred between the gene pools of two extant organisms.

Hemoglobin

Two molecules—proteins and nucleic acids (DNA)—are used for this kind of analysis. These are the molecules characteristic of living matter that govern the functions needed to live. Let us start with proteins, since that was the system first used to establish the connection between humans and apes. We will discuss DNA in more detail later. The body holds tens of thousands of proteins. One of the most well known is hemoglobin, which is the main constituent of red cells in the blood. Red cells are spheres about eight-thousandths of a millimeter across which form roughly half the blood. Hemoglobin gives blood not only its red color but also its most important property—the capacity to collect

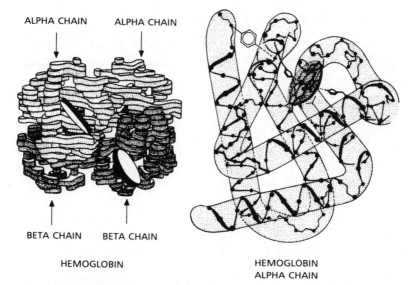

ALPHA CHAIN   ALPHA CHAIN

BETA CHAIN   BETA CHAIN

HEMOGLOBIN

HEMOGLOBIN
ALPHA CHAIN

2.2 A hemoglobin molecule is made of four globins; a pair of these are also known as alpha chains and a pair as beta chains. *Left:* The four chains are shown as solid figures and in their relative proportions. *Right:* A typical chain that appears like a spiral string of about 150 pearls formed by its constituent amino acids. The disks are heme molecules.

oxygen from the lungs and deliver it to body tissue for use in cells. All cells obtain much of their energy by burning oxygen.

A hemoglobin molecule is made of four submolecules—four chains that is, held together in a special way. There are two types: two identical chains, known as alpha; and two others, also identical but different from the first, called beta. Each has a smaller molecule within, which stores the oxygen and permits its exchange with the tissue. This smaller unit called *heme*, contains iron; many anemias are treated with iron because hemoglobin has been lost and new hemoglobin needs to be produced.

The four chains are fundamental for function: on its own, iron, or the heme that contains it, could not exchange oxygen. The chains are linear and look like rows of pearls. There are 141 "pearls" in an alpha chain and 146 in a beta chain, and 20 different types of pearl.

These units ("pearls") are known as amino acids, each of which has a name (alanine, glycine, tryptophan, serine, and so on). They determine the properties of the protein. In defining a protein's functions, we need to know which amino acids it contains and, even more important, their sequence.

Studying Evolution Through Proteins

A protein tends to be virtually identical in all members of the same species, but is different in organisms separated by evolution, because a number of amino acids have been added, substituted, or lost. In general, the greater the differences between two organisms, the greater the difference between the constituent amino acids of a protein.

Taking the start of the alpha hemoglobin chain in three fairly different species (human, horse and chicken), and using letters for the various amino acids (V for valine, L for leucine, S for serine, and P for proline, and so on), we obtain the following sequences:

human	VLSPA	DKTNV	KAAWG ...
horse	VLSAA	DKTNV	KAAWS ...
chicken	VLSNA	DKNNV	KGIFT ...

fifteenth) are different from the horse's, and six out of fifteen are different from the chicken's. There are six differences between the horse and chicken, too.

This shows us something of interest: the general structure of the hemoglobin is similar in all three, despite the considerable difference between the species themselves. The most similar organisms (human and horse), however, have only two different amino acids, whereas the more radically different chicken has six different amino acids. This tallies with the fact that the chicken is a bird, but horses and humans are both mammals and therefore more closely related.

How can this distance be measured in terms of evolution? Today we believe that mammals have existed for at least sixty-five million years, and started to spread only after a cataclysm that occurred around that time. According to prevailing thought, a huge meteorite probably fell in an area of today's Gulf of Mexico, creating a colossal dust cloud that eclipsed the sun, leading to many years of winter, which profoundly changed the climate. Large animals such as the dinosaurs died out, while small, warm-blooded animals like the mammals of the time survived and began to proliferate. That mammals started to diversify around sixty-five million years ago is therefore plausible. We know from fossils that birds began to diversify around two hundred million years ago.

There is a ratio of 1 : 3 between the difference of two amino acids between humans and horses (out of the fifteen given above) and that of six between humans and chickens (or horses and chickens). This is roughly the same ratio as the time taken for humans and horses (sixty-five million years), and birds and mammals (two hundred million years) to evolve separately. Using many proteins and species, scientists have verified this simple relationship between the divergences found in the biological makeup of species and the time they take to accumulate. It is the basis of the so-called molecular clock, used to calculate evolutionary divergence times on the basis of differences between species, and to reconstruct evolutionary family trees.

Obviously, one statistic using fifteen amino acids is not enough. If we look at the number of differences in the whole hemoglobin alpha chain between four mammals, the following picture emerges:

Number of differences in alpha chain hemoglobin amino acids in humans, gorillas, pigs, and rabbits

	Human	Gorilla	Pig	Rabbit
Human	0	1	19	26
Gorilla	1	0	20	27
Pig	19	20	0	27
Rabbit	26	27	27	0

Our own eyes tell us that humans and gorillas are more alike than humans and pigs, but it is difficult to decide just by looking whether pigs or rabbits are closer to humans. The table shows that it is the pig, because there are fewer differences in amino acids. But more data would be necessary to reach a safe conclusion.

Humans' and Apes' Evolutionary Family Tree

At this point we can begin to create an evolutionary family tree. First we group humans and gorillas together, because they only have one amino acid difference. Humans and gorillas respectively have nineteen and twenty differences when compared with pigs, which persuades us that pigs broke away much earlier than humans and gorillas—we can now add a branch for the pig. Finally, we can put the rabbit on the outermost branch because it has twenty-six differences from human, and twenty-seven from both gorilla and pig. Obviously the longer the chains analyzed, the better, because greater numbers have greater statistical value. Best of all is to compare many proteins and DNAs.

These statistics can also be obtained using a less direct method for calculating the number of differences in amino acids, which is how two scientists from the University of California at Berkeley—Vincent Sarich,

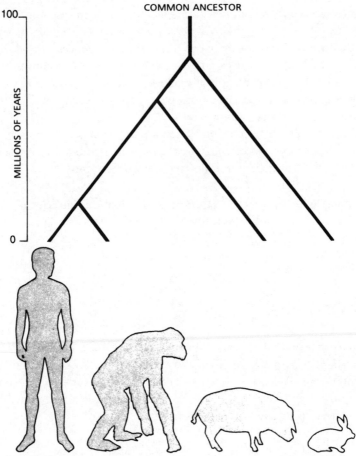

2.3 Evolutionary tree of four mammals reconstructed on the basis of the hemoglobin alpha chain.

anthropologist, and Allan Wilson, a biochemist—were able to place the split between humans and apes at five to seven million years ago.

This conclusion was originally clamorous, because the split had previously been thought to occur much earlier (about twenty to twenty-five million years ago, on the basis of not very dependable paleontological data). Lucy was subsequently found, with her numerous apelike features, but she was also clearly similar to humans. She therefore must have lived after the separation of the two, but not by much. We have said that Lucy can be dated fairly reliably as 3.2 million years old, which obviously tallies better with the five to seven million years suggested by molecular biologists than twenty to twenty-five million years.

It is now commonly thought that humans and chimpanzees separated around five million years ago—some people say five to seven million, others say four to six million. The split with gorillas occurred earlier, and that between humans and orangutans earlier still, about ten to fifteen million years ago.

Genus and Species

What does it mean to belong to the same genus or the same species? The system most commonly used was invented in the eighteenth century by the first and greatest classifier of living organisms, Carolus Linnaeus of Sweden.

A living creature is distinguished by two Latin names. The first is the *genus*, or group of similar species, and the second the *species*, meaning the group of individuals capable of producing fertile offspring. For example, donkeys and horses belong to the same genus (*Equus*) but not to the same species. Hybrids, known as mules and hinnies, are sterile.

All living humans belong to the same species, *Homo sapiens*. There are no substantiated reports of attempts to crossbreed them with apes. If such experiments have been conducted (perhaps by Nazi pseudoscientists), one hopes they will never be repeated. Where would the eventual offspring live, among us or in a zoological garden? The very thought should be enough to dissuade any sensible person from trying such an experiment.

The Oldest Ancestor

The australopithecines are the earliest humans after the ancestor we share with the apes. They are no longer protoapes, even if they are not yet, strictly speaking, humans.

We know there were several australopithecines; *afarensis*, *africanus*, *aethiopicus*, *robustus*, and *boisei*. It seems that among these the real missing link was *Australopithecus afarensis*, who had two main types of descendant: human beings, and a couple of australopithecine species, who inhabited Africa until about one million years ago, before dying out. But a very recent discovery, *A. ramidus*, 4.4 million years old and also from Ethiopia, may take *afarensis'* place as a missing link.

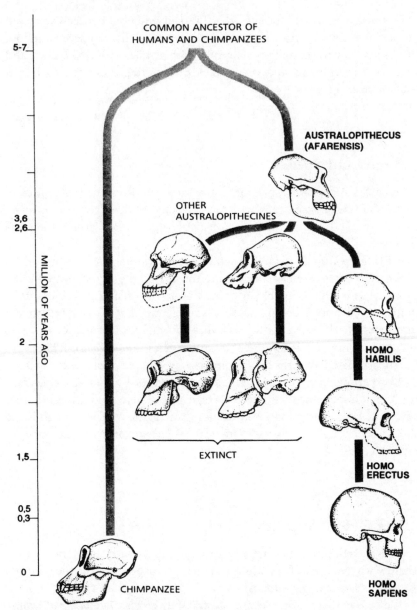

2.4 Australopithecine family tree, including the genus *Homo*, giving approximate dates in which each lived. Based on fossilized skulls found in eastern and southern Africa.

The Human Tree

At least three species of the genus *Homo* have been distinguished: *Homo habilis*, the first and oldest, from around two to two and a half million years ago; *Homo erectus*, found roughly two million to about three hundred thousand years ago; and, lastly, *Homo sapiens*. Only *Homo sapiens* sometimes has additional subspecies to better classify the different types known.

The definition and classification of fossil species like *Homo habilis* and *Homo erectus* was not always agreed upon. The same is true of most of the enormous number of plants and animals that have been classified. With fossilized species, in particular, there is simply no hope of obtaining information through crossbreeding, since the definition of belonging to a species implicates the possibility of absolutely controlling the ability to have fertile offspring.

In the taxonomists' defense, it should be said that their guesses have often proved correct. Where it has been possible to check their assumptions (for organisms other than humans), they have usually been right.

The Australopithecines

The australopithecines' cranium was small in relation to their face, and as a result they presumably resembled apes. The internal volume of the cranium, which corresponds more or less to the volume of the brain, is about 400 cubic centimeters, not much more than a chimpanzee or gorilla, which are rather bigger animals. Lucy is only 3′ 7″ tall. Her cranium not only is small but also is low down. This feature changes radically only with modern humans, whose cranium and forehead are both higher. Lucy has a protruding face and heavy jawline. From the shape of the junctures between tendons and muscles and bone, we can tell that she was a powerful creature. Her teeth are already changing in a way that will later be more pronounced: apes' canines are highly developed and separated from the incisors. Human canines, however, are no longer important, and the gap with the incisors has closed. *Australopithecus afarensis* has medium-size canines, but they are separated from the incisors to the same degree as an ape's.

The australopithecines were erect, even if they stood slightly differently than we do. Both chimpanzees and gorillas stand, but they normally walk on all fours using the knuckles of their hands in a way totally unlike us.

Many species of australopithecine have died out, the last disappearing about one million years ago. Still, in East Africa, the differentiation of the animal that would be the first human was beginning. Louis Leakey found it again in the Olduvai Gorge in Tanzania.

Homo habilis

What makes us consider *Homo habilis* the first of our species? The braincase has grown (from 400 to 630 cubic centimeters in the million or so years separating this creature and the earliest australopithecines we know of). Fashioned stone tools also appear in place of the very rough pieces of stone found previously. (An exhibition of both bones and tools has been opened on the spot where Leakey made his discovery.) This human is erect but is three and a half feet tall and like Lucy, has long arms, which may have been needed to climb trees with the same agility as the apes.

Dating from two to two and a half million years ago, *Homo habilis* (which means "skilled worker, handyman") had the ability to use tools. It has recently been claimed that some of the australopithecines may already have known how to make stone tools.

Homo habilis's tools are rough. Many are just stone chips: small ones were used as scrapers and larger ones as axes and choppers. Some have been found alongside animal bones; presumably they were used for stripping meat or breaking the bones to get at the marrow. The technique used for making these tools is *Oldowan*, from the place (Olduvai) in which they were found.

The discovery of abundant stone deposits and bones tells us that this human was a scavenger, and a meat eater (today's chimpanzees are mostly vegetarian).

Homo erectus

The human who followed, *Homo erectus*, lived from two million to approximately 300,000 years ago. We know little of the transition between *habilis* and *erectus*, because there are few well-dated and researched fossil or archaeological remains. The change from *Homo habilis* to *Homo erectus* is marked by further growth of the braincase, which, perhaps very gradually, increases from an average of 630 cubic centimeters to a little over 1,000, while tools become more common, specialized, and better made.

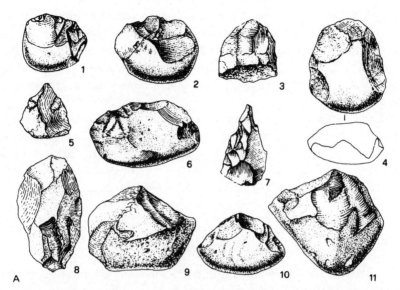

A

1, 2, 3, 4, 6, 7, 9, 10, 11, choppers; 3, 8, strippers; 5, retouched stone flake.

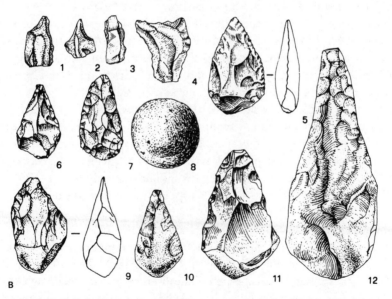

B

1, 2, 3, 5, 6, 7, 9, 11, stone chips; 4, scraper; 8, bola; 10, 12, bifacial cutters.

2.5 (A). *Homo habilis*'s tools (Oldowan culture). (B) *Homo erectus*'s tools (Acheulian culture). Scale: 1/4 actual size.

With the growth of the larger braincase, the cranium moved upward, so to speak, although the browbone, with its bony ridges, remained very pronounced. The face shortened, but the teeth continued to protrude. The jawbone was shorter but remained turned under, a feature that disappeared only with the modern human, who has a fully formed, pointed chin.

The brain growth was probably connected in part to the development of tools. The new tools are called *Acheulean* (from the Saint Acheuls site in France). They can be found up to two or three hundred thousand years ago, on sites inhabited by *erectus* successors. The twin-headed (bifacial) axe came into use and remains for a long time.

The Places of Human Evolution

As Darwin and Huxley had intuitively foreseen, for more than several million years Africa was the theater of human prehistory. Their reasoning was simple: the creatures closest to humans (chimpanzees and gorillas) today live mainly in Africa, so humans must have originated there.

From what we can gather, our human ancestors—and the australopithecines before them—had already left the forest (where the apes still live) and inhabited areas similar to the modern savanna: tropical grasslands with tall vegetation, bushes, and scattered trees, inhabited by large quadrupeds and a wide variety of plant and animal species.

About two million years ago, *Homo erectus* began to travel and colonize Asia, Europe, and virtually all the Old World during a period of hundreds of thousands of years. This expansion probably was possible because tools were more evolved and intelligence greater. (*Erectus* brain was more than twice the size of an australopithecine's).

The first place we know they stopped is in the Middle East, at Ubeidiya in today's Israel. Remains dating back a little more than one million years have been found here, and are the first known signs of human presence outside Africa. Subsequently, *erectus* reached Southeast Asia, Java, and the Far East (with Peking Man). The story of the well-known Peking Man is strange. The finds were first made at Chou-kou-tien (now written Zhoukoudian). When war between Japan and the United States became likely, the American embassy in Peking helped organize a train protected by marines to take the *Homo erectus* skulls to safety in the United States. The train—which left China on the same day as the Japanese surprise attack on Pearl Harbor, precipitating the U.S.

declaration of war—was intercepted by the Japanese, and the archaeological material vanished. Fortunately, a German paleoanthropologist,
Franz Weidenreich, had taken some excellent plaster casts and highly
detailed notes on the remains.

Chinese archaeologists later continued excavations, and today the
collection numbers 100,000 pieces. There were once thought to be
significant differences between the Acheulean tools made in the East and
in the West, and that this indicated profoundly different cultures. We
now think the differences have been overstated. In East and Southeast
Asia, there are relatively few twin-headed axes. It has been suggested that
the widely available bamboo plant was used here for many jobs that
elsewhere were done with stone tools.

Peking Man has left us one of the clearest examples of the use of fire.
This example came quite late (about three hundred thousand years ago),
but it is plausible that fire was known much earlier, and that *erectus*
spread partly because of the ability to make it; perhaps they took this
utility with them from the start of their travels. This interesting idea is
only hypothetical so far.

Homo sapiens

The first remains of our species have been dated between three and
five hundred thousand years ago. *Sapiens* may have existed alongside
Homo erectus for a while, but we cannot be sure. With archaic *Homo
sapiens* the braincase quickly reached its final capacity of about fourteen
hundred cubic centimeters on average. This has remained much the same
ever since, with a slight difference between men and women (most
probably due to the difference in average weights and statures) but with
enormous individual variations. In modern human populations, brain
volume has little relation to actual intelligence: one can have a small head
and be very intelligent, or vice versa. During human evolution we see a
dramatic growth in brain volume, most probably linked to improvements in certain intellectual capabilities, such as understanding how to
make tools and use language in an increasingly complex manner. *Homo
sapiens* seems to have always had much the same braincase size. The
shape, however, has changed, and for this reason the *Homo sapiens* of
three hundred thousand years ago is known as archaic *Homo sapiens*. In
archaic skulls, the heavy, arched brow ridges and thick bones remain,
while the face still protrudes somewhat like an ape's.

The Differentiation of *Homo Sapiens*

Pictures were recently published of a cranium that may have belonged to an intermediate human, between *erectus* and *sapiens*. Found in China, it has been dated at around four to five hundred thousand years ago. This find is very interesting, although we won't know its real significance until other data have supported it.

Fossils and archaeological sites become steadily more numerous the closer we come to our own times, and our information is a little less nebulous, although the question marks are still plentiful. Archaic *Homo sapiens* is found in Africa (North, South, and East), throughout Europe (except Scandinavia), and in Asia, particularly South and Southeast, but also western (in other parts less research has been done). It becomes harder to consider *Homo sapiens* a single entity. It was difficult for *Homo erectus*, for whom we have fewer clues, but now both spatial and temporal differences become clear.

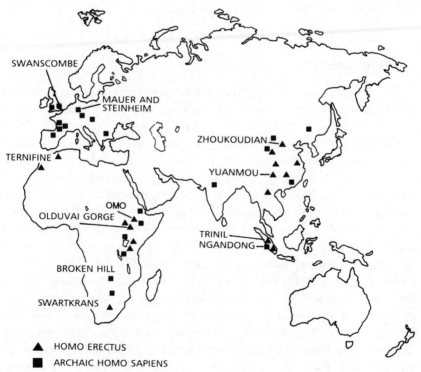

2.6 Sites of major *Homo erectus* (triangles) and archaic *Homo sapiens* (squares) fossil finds.

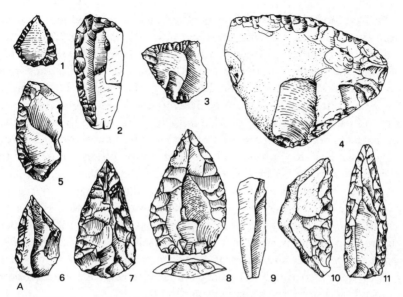

1, 6, 7, 8, tips; 2, 3, 4, 5, 10, scrapers; 9, 11, blades.

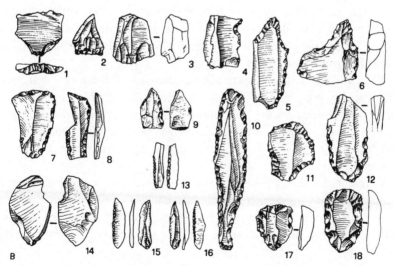

1, stone chip; 2, 3, cores; 4, 10, blades; 5, 6, 7, 11, scourers; 8, 9, 12, 14, burin; 13, 15, 16, blade flakes; 17, 18, keel-bottomed strippers.

2.7 *Homo sapiens*: (A) Mousterian and (B) Aurignacian tools (1/2 actual size).

We have noted the distinction between archaic and modern humans. From two hundred thousand years ago onward, European *Homo sapiens* is different from that in Africa or China. These *Homo sapiens neanderthalensis* are among the last of our ancestors. Around sixty thousand years ago, they also turn up outside Europe, in the Middle East, only to disappear completely about thirty-five thousand years ago. Archaeologists and paleontologists are divided in their views. Some think they died out, partially because of pressure from modern humans who spread from the Middle East to Europe and replaced them; others think that the Neandertal actually became modern humans. If they died out, we must consider them cousins, or uncles, who died without progeny; but if not, they are the direct ancestor at least of the European peoples. I favor the first theory. In the last one hundred thousand years, then, modern humans appear, indistinguishable from ourselves and known as *Homo sapiens sapiens*. They are different from the earlier *sapiens* in important ways that are increasingly relevant to ourselves: language is much more advanced, and stone tools change significantly.

Toward the end of the archaic *sapiens* period, the old Acheulean method of toolmaking, which had been used for more than a million years, began to be replaced. Different-shaped stones were chiseled into various tools, then repeatedly touched up to improve their efficiency. The new tools are called *Mousterian*, after the cave at Le Moustier in the French Dordogne, where they were first found. (This site is near Cro-Magnon but is substantially older).

When and where the Mousterian style was adopted are not clear, but it quite quickly replaces the old one and generates a wide range of more varied and refined tools. It continues to be the predominant style in Europe, Africa, and the Middle East until forty or fifty thousand years ago, when the even more evolved Aurignacian style (from the French Aurignac site) appears.

At that time, in Europe at least, there were two separate peoples: Neandertals and modern humans, different not only in head shape but also in their tools (respectively, Mousterian and Aurignacian). Mousterian tools disappear with Neandertals. We are entering the modern age.

THREE

One Hundred Thousand Years

By three hundred thousand years ago, the human brain had virtually reached its present size. It may even have been a little bigger in volume. This does not mean its internal structure was the same as ours. The fact that tools remain rough for a long time to come suggests that significant change may have been necessary for the brain to achieve its present standard of perfection. In the next one to two hundred thousand years, the most primitive facial features disappear (archaic *sapiens* is still rather apelike in appearance), and tools evolve considerably. The new Mousterian technique we associate with *Homo sapiens* takes over from the old Acheulean one. There is a much wider range of these tools, and they show signs of careful retouching and maintenance. When and where the Mousterian style and *Homo sapiens* first appear, however, remain a mystery to us.

What we can be fairly sure of is that in the last three hundred thousand years a special kind of *Homo sapiens* developed in Europe—the Neandertals, who assumed their most representative features some sixty thousand years ago. In Africa, during the same period, we find an archaic *sapiens*, who looks more like a modern human. These facts tally with a number of others that point to Africa as the modern human's birthplace and center of their early movements, although the idea is not embraced by all experts. Passionate debate continues, and no easy resolution of the question is in sight. A number of paleoanthropologists attribute great importance to human remains found in China, which some—but by no means all—observers think resemble the modern Chinese. Their suggestion is that modern humans emerged not in one single area, but across a huge zone covering practically the entire globe. The two opposing theories, an "African" and a "polycentric" development, are the subject of intense anthropological discussion. From a geneticist's point of view, a single origin followed by expansion is the more credible of the two.

Nearly all our most important fossil finds have come from nonarchaeological sites in various continents. The richest sources of material have been road, rail, and housing construction works. Europe has a higher population density than Africa and other parts of the world, except for China, Japan, and India (where, however, archaeological research is now actively being done). As a result, more of Europe has been unearthed for building purposes, and when interesting material has come to light, it has often found its way into museums or universities.

This explains Europe's preeminence to date in terms of both finds and depth of paleoanthropological analysis. Both Neandertal and Cro-Magnon discoveries have been made; this is logical for the Neandertals since they lived mainly in Europe. Cro-Magnon, however, is a fairly late example of the modern human. Older examples have been found elsewhere who, like the Cro-Magnon people, are almost indistinguishable from the humans populating the earth today.

Homo sapiens neanderthalensis

The Neandertals have aroused much curiosity because their features are clearly very primitive, and sufficiently different from ours to make us treat them as a thing apart. A special category within Homo sapiens—Homo sapiens neanderthalensis—has been created for them. Some treat them as a subspecies of Homo erectus, or even as a separate species altogether. A quantitative analysis by Professor W. W. Howells of Harvard University shows there is no particular resemblance between Neandertal skulls and modern European ones or those of any other surviving human race.

I am grateful to J. J. Hublin of the Paris Museum of Man for kindly showing me two famous original skulls side by side, a Cro-Magnon one that clearly belongs to a modern human and a Neandertal one from La Chapelle aux Saints in France. The difference is astonishing, although it should be said that Neandertals vary considerably and other skulls from eastern Europe are perhaps less extreme.

The Neandertal's braincase is the same size as a modern human's and sometimes even a little bigger. However, the skullcap is longer and lower, making the forehead considerably shorter. The face is long, and has a wide nose and weak chin. The pointed chin worn by nearly everyone today appeared only with modern humans. There is also a kind of bony chignon at the back of the skull. The bones show that these people were very strong.

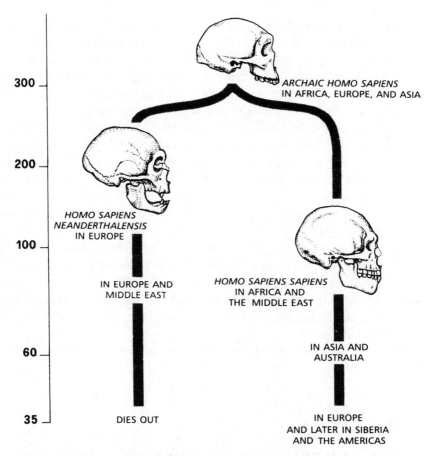

3.1 Family tree of *Homo sapiens* with comparison of Neandertal and modern humans. The scale indicates thousands of years ago.

A strange feature, particularly noticeable among the elderly, is pronounced wear on the outer edge of the incisors. This is clearly not inborn, and would seem to be caused by use of the incisors to grip materials, maybe cord and so on, when making tools. This type of wear is still found among the Eskimos, albeit to a much lesser extent.

The Life of a Neandertal

The Neandertals adapted to life in varying climates and survived through several Ice Ages. We think they lived in caves, because this is where nearly all the most important finds have been made. Remember, however, that archaeologists like digging in caves because there is a better chance of finding well-preserved, almost undisturbed relics. Open-air sites have been found, often close to water sources, but unlike modern human settlements these show few signs of modification in preparation for extended usage. Open-air Neandertal sites have yielded mainly stone remains—perhaps they were used as tooling sites or places to cut up game after the hunt. Neandertals were undoubtedly hunter-nomads who perhaps made forays from central camps, and meat—venison, cow, ox, and horse—was their staple food. Mousterian tools are always found in their dwellings, making them an excellent indicator of Neandertal presence in Europe, although it should be remembered that the Mousterian style lasted a long time and was used by archaic *sapiens* and modern humans as well as the Neandertal.

In addition to knives, scrapers, punches, and many other tools, spears have been found, including a wooden one still embedded in an elephant's side. Obviously, the hunt went well that day. Their most important materials were stone and wood. They did not use bone or ivory at all, and in marked contrast to their successors, they left us no signs of interest in art.

The discovery of skeletons showing the marks of age or long illness suggests they cared for the elderly and sick.

Ritual Behavior

Neandertal burial places have been found in caves. The body generally is placed in a crouching position, although this may simply be in order to dig a smaller hole and may have no ritual significance.

Mounds of broken bones have been found, suggesting that Neandertals ate the brain and bone marrow. It does not imply that they killed their neighbors in order to eat them. They may have been necrophagous, meaning they ate their dead, a form of cannibalism still widespread in Africa and practiced in New Guinea until a few years ago.

When I was working in the Central African Republic, there were occasional outbreaks of anthropophagy, even though it is forbidden and punishable by law. Apparently, Bokassa himself, the self-styled president of the Central African Republic, feasted on the bodies of dissenting students murdered by his police. Necrophagy has accompanied humanity throughout history. In 1967 I visited an area of New Guinea where necrophagy was still customary. It was later found to cause *kuru*, a serious neurological syndrome that is inevitably fatal within a few years.

At the time kuru was thought to be a genetic disorder, because it was concentrated within certain families, but it eventually turned out to be a virus transmitted when children eat their dead parents. Even the preparation of the body for funeral rites, generally by the women, can lead to infection.

The Neandertal bones found so far give no indication of injuries deriving from human attack. There are various lesions—teethmarks and so on—apparently caused by carnivorous animals. Other marks indicate that the bones were stripped of flesh, perhaps before burial. Burial within the cave may have developed as a way of avoiding predators, and the habit of stripping the bones perhaps grew to avoid the smell of decomposition.

Other finds suggest that burial rites may have taken place. During foundation work for a villa at Monte Circeo to the south of Rome, a circle of stones arranged around a skull was discovered on a cave floor. Unfortunately, the discoverers removed the skull before a team of anthropologists could arrive, and the original position of the head remains unknown. The base of the skull had been opened, maybe in order to eat the contents. Further research has not solved this mystery.

A great deal of perhaps undeserved noise has been made about the discovery of flower pollen in another Neandertal grave. Some people thought that flowers had been placed there in remembrance, but this is very uncertain and the material may well be intrusive. Establishing precise sources is one of the biggest problems in all excavation work. To conclude, we aren't sure if there was any religious intent in burial, and many archaeologists actually doubt it.

Neandertal Presence

Neandertal finds have been made throughout eastern and central Europe, and as far afield as the Middle East, and east of the Caspian Sea.

We have been unable so far to resolve the question of presence in different eras. We find modern humans in the Middle East very early on —one hundred thousand years ago. The same areas have yielded examples of later Neandertal presence. In Israel, Neandertal and modern human remains were found in caves in the same area and dated using radiocarbon at forty thousand years old—the maximum age that radiocarbon can test to. Subsequent, more accurate dating has shown that the two groups lived earlier and at different times. There is a theory that modern humans unsuccessfully attempted to settle the Middle East

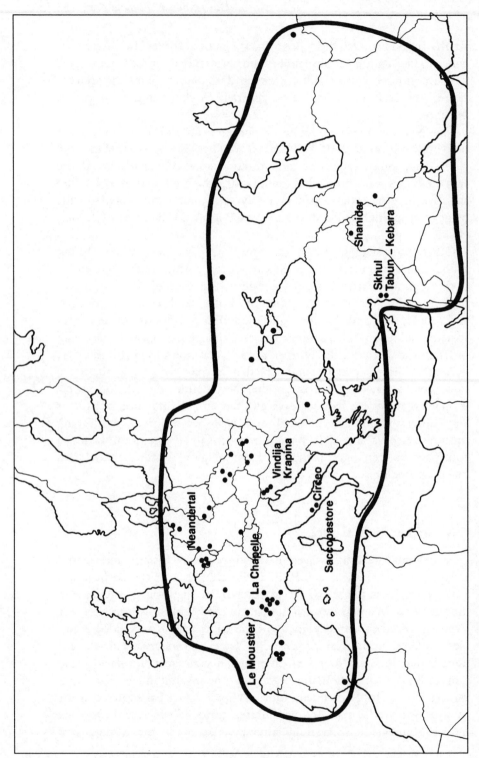

3.2 Maximum spread of Neandertal settlements.

around one hundred thousand years ago, and that this was followed by a period of Neandertal occupation, which at some point ceased, after which modern humans returned.

The last traces of Neandertal people are found in Europe, dating back around thirty-five thousand years.

Homo sapiens sapiens

The humans we call "*Homo sapiens sapiens,*" or also "anatomically modern humans," are those whose skeletons we cannot distinguish from ours. They appear late (around 40,000 years ago), in Europe, where they apparently replace Neandertal, but there are earlier finds of modern humans in Africa and in the Middle East. The earliest evidence left by our direct forebears in Africa comes from two finds of almost the same age made in the South African Border Caves and at Klasies River Mouth. These sites have been roughly dated at 74,000 to 130,000 and 74,000 to 115,000 years ago, respectively. They reinforce the view that modern humans originated in Africa, as does the fact that the archaic *Homo sapiens* who lived in various parts of Africa more closely resembles modern humans than the archaic *Homo sapiens* found elsewhere in the world.

In the latter half of the 1980s, archaeologists found and accurately dated various modern human sites in the Middle East. The first and perhaps most important is at Qafzeh, in Israel. Its date has been set at 109,000 to 92,000 years ago, although both dates are subject to statistical error. This is much the same age as the African and Middle Eastern finds.

Two New Dating Systems

The Qafzeh site has been dated using two new systems, *thermoluminescence* and a similar one, *electron spin resonance* (ESR).

Thermoluminescence is particularly useful for dating fired clay and materials that were subjected to high temperature on manufacture. It was used at the Qafzeh site to date stone instruments that were almost certainly fired at the time they were made. The method is based on the fact that uranium, radiopotassium, and other radioactive elements leave transmutation traces as they decay. These traces can be counted because they generate visible light when heated. However, it is important to count only the traces formed after manufacture, and not the ones previously present in the matter. Firing has the effect of zeroing the clock, as it were, by removing all trace of previous transmutations. This explains the particular usefulness of thermoluminescence for fired clay.

The Spread of Modern Humans

To simplify matters a little, here is the chronology of our most recent ancestor (*Homo sapiens*): Three hundred thousand and perhaps more years ago various types of archaic *sapiens* already peopled various parts of the world. Two hundred thousand years ago, Neandertal can be found in Europe, and one hundred thousand years ago modern humans (*sapiens sapiens*) are found in South Africa and Israel. Subsequently, Neandertal appears in the Middle East about sixty thousand years ago, when there is no sign of modern humans in the area.

Homo sapiens sapiens then starts to appear everywhere. Within sixty to seventy thousand years, the species reaches every corner of the globe, manifesting an ability to adapt to the most varied environments and also —let it be said—possessing a strong spirit of adventure.

In China, there is a *sapiens sapiens* relic more than sixty thousand years old. Unfortunately, it is the only one we have and its date is somewhat uncertain. Modern humans seem to have reached Australia and New Guinea during this time, and to do so they must have used seaworthy vessels. The stretch of water dividing these islands from southeast Asia was shorter than it is today, but without some kind of watercraft it still would have represented an insurmountable obstacle. In Australia, a fossil that is generally accepted as a modern human dating back thirty-five to forty thousand years has been found and also sites from fifty-five to sixty thousand years ago. Modern humans came late to Europe. The earliest traces we have are from eastern Europe, around thirty-five to forty thousand years ago. The chronological pattern of finds suggests they came from the east. In much the same period, we find the last evidence of Neandertal presence.

Modern humans subsequently moved into the colder regions of Asia. This was undoubtedly an extremely difficult conquest, because Siberia is one of the coldest places on earth. Cultural and very probably biological adaption, too, were needed to survive in this climate. From Siberia they journeyed to America, at the latest fifteen thousand years ago, but perhaps earlier, presumably taking advantage of the fact that the shallow waters of the Bering Strait became dry land during the last Ice Age when part of the ocean's waters was absorbed by huge polar glaciers.

Modern Humans and Neandertal: Friction or Fusion in a Common Environment

It is extremely hard to say whether modern humans supplanted the Neandertal, perhaps eliminating them physically, or whether they merged, not least because we have no way of knowing if they were the same species or two different ones. Generally speaking, where two different species compete for the same resources in the same environment, eventually only one survives. Fusion cannot be ruled out, but at the time that they both lived in Europe there were important cultural differences, which makes the idea less plausible.

When biologically modern humans appeared—in South Africa and Israel about one hundred thousand years ago—they used Mousterian-style tools, as did the Neandertal and all archaic *sapiens*, whether in Africa or elsewhere. This mystery is yet to be resolved. It seems possible that human biology changed before its culture, and that cultural evolution was slower to occur. The opposite is usually true: cultural change is faster than biological evolution. If, however, there is a biological basis to cultural change, such as the development of new cerebral structures facilitating cultural evolution, there may be a long period of biological evolution at the end of which cultural evolution is rapid. Perhaps modern language developed in a similar way during this period.

We have already seen that around fifty thousand years ago modern humans stopped using Mousterian tools and adopted a new style known by various names, including Aurignacian, and took it with them to Europe, where it is so different that it is used to distinguish modern human sites from Neandertal ones in the absence of human remains. This is a little risky, but so far the association in Europe of Neandertal with Mousterian and modern humans with Aurignacian has proved reliable.

The range of Aurignacian tools is wider and more varied than the Mousterian one. There are many types of instrument with precise shapes and recognizable uses, implying a higher degree of specialization. There are tapering stone chip blades with fine edges and very sharp cutters and scrapers. Ivory, horn, and bone are employed as well as flint.

The Aurignacian technique spread very rapidly and generated local cultures with different tools. Classification of these has led to an explosion in archaeological name tags. Glynn Isaac, a very fine archaeologist who unfortunately died a few years ago, noted that this great variety coincides with the beginnings of major language diversification. Modern human language, in fact, now started to reach a high level. Neandertal language is thought to have been less developed than that of modern humans; and some experts believe that the Neandertal pharynx-larynx was not long enough to permit the same vocal range.

The Mousterian style varied from place to place, but nowhere near as much as the Aurignacian. The great local diversification of tools coincides with the spread of humans. Linguistic diversification was probably simultaneous and took place for the same reasons: independent evolution in communities that were cut off from one another.

A New Lifestyle

In contrast to Neandertal, modern humans display an enormous interest in art. This behavioral difference is, in my opinion, another major reason for believing that the two human types were significantly different. Everywhere we find rock drawings and other marks. The first manifestations came from the southwest of France near the Pyrenees and undoubtedly accompanied the arrival of modern humans there thirty-five thousand years ago. Art suddenly flourishes, particularly in the area between today's Dordogne and the Franco-Hispanic Atlantic coast (see Figure 9.2), where an enormous number of decorated caves (150) have been found. This was an excellent place to live—and still is today. Together with cave paintings that depict mainly animal forms with extraordinary precision and expressiveness, we find rock carvings and stone statuettes (frequently of women with enormous stomachs, breasts, and hips). Complex techniques for working animal teeth, shells, bone, ivory, and soft stone were used for personal objects and ornaments.

An important innovation is the bow, which appeared some twenty-thousand years ago and spread rapidly. The spear was made to travel faster and farther by fixing a thrower in the form of a short extension stick to the base of the shaft. This type of spear with a very hard obsidian tip, known as a Clovis Point, was used to hunt mammoth in America. More refined hunting equipment and the first fishing tools (harpoons and hooks) appeared.

Archaeological sites show that both settlement type and distribution differ from Neandertal ones and are similar to those of today's hunter-gatherers.

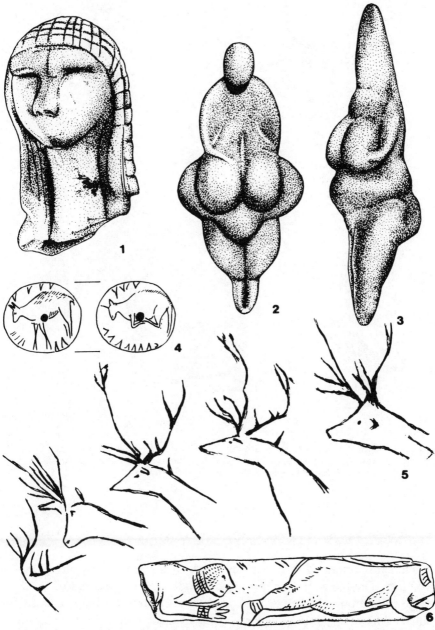

3.3 Upper Paleolithic wall painting and artifacts from Europe: (1) 1.5″ statuette head from Brassempouy in France (ca. 22,000 years old); (2) 6″ engraved mammoth ivory statuette of a woman, from Lespugue in France; (3) Savignano Venus (Emilia Romagna, Italy); (4) front and reverse of tiny engraved bone disk with hole through the center from the Dordogne in France; (5) rock painting of wading deer from the Lascaux Cave in France; (6) bone engraving of woman being chased by man, Isturiz Cave in France.

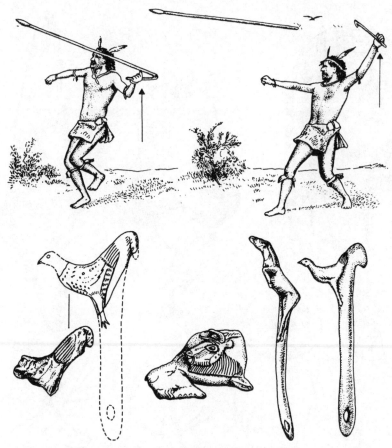

3.4 Spear-thrower used as an arm extension to make the spear travel farther and faster. It was very common in the Upper Paleolithic era, from Europe to Australia and America. *Below*: Examples of decorated spear-throwers.

Humans continued to live in caves, which they frequently altered for comfort. The remains of skin-covered wooden frames weighed down by large animal bones have been found, telling us they also built tents and huts. Some of these are quite large. They made clothes from skin and fur and invented a needle to sew them together. Three occupants of a burial site to the north of Moscow show signs of complete clothing, including hoods, shirt, jacket, legwear, and moccasins.

3.5 Upper Paleolithic cave decoration in Europe: 20 × 10 yard drawing of oxen found on the walls of the Altamira caves in northeast Spain.

The Question of "African Eve"

Much has been said in recent years about "African Eve," a name that is striking, but also misleading for a number of reasons. The name seems to imply that the entire human race somehow descends from the same original "mother." Not surprisingly, this idea pleases Christian fundamentalists, since it appears to support a literal interpretation of the Bible (as long as we overlook the fact that, based on molecular data, the so-called Eve was born about two hundred thousand years ago, and not six thousand as the Bible says). A brief aside on biology is needed before we explain this theory, which is the result of laboratory research on mitochondria by Allan Wilson, the excellent biochemist from the University of California at Berkeley, who turned his talents to the study of evolution. Unfortunately, he died prematurely in the summer of 1991.

The mitochondria are small organs found in the cells of all higher organisms. They use the oxygen introduced by breathing to produce energy and act a little like a power station. They are not the cell's only way of producing energy, but are the most efficient. Mitochondria are found outside the cell's nucleus, in the liquid between nucleus and outer membrane. There may be dozens or even thousands in any one cell, but there will be at least one in almost all cells. They resemble a small bacterium in shape, and most probably originate from a bacterium that over a billion years ago adapted to live in symbiosis with the cell and became extremely important by assuming the role of power station.

The mitochondrion is, however, to some extent independent of the rest of the cell, since it has its own tiny chromosome, which, like all chromosomes, is made of DNA. DNA is the structure that carries inherited characters—it contains all the hereditary information needed to transform inanimate matter into living matter and so to build new organisms. DNA directs the transformation of nutriment into new cells —children or descendants, in other words. It is worth taking a closer look at DNA, because it is the substance that allows biological features to be handed down from generation to generation in every single living organism.

The Structure of DNA

DNA, composed of very long strands, forms the famous chromosomes, which are segments of DNA present within the nucleus of every cell. The chromosomes are the same in every cell and distinguish each individual from everyone else. This does not mean you can see that one person's chromosomes differ from another's simply by looking through a microscope. The difference is much finer because chromosomes are enormously detailed.

A chromosome is made up of a huge number of units, of which, however, there are only four kinds: A, G, C, and T. These are the initials of four simple and well-known chemical compounds (adenine, guanine, cytosine, and thymine). A and G belong to the purine group of substances, which also includes caffeine and uric acid; C and T are pyrimidine and have smaller molecules (vitamin B1 derives from a pyrimidine). Chemically, they all behave as alkalies or bases (the opposite of acids), so they are often known simply as bases. Each base is attached to a sugar molecule, the sugar being deoxyribose (hence the *D* in *DNA*). The structure of a DNA strand is actually very simple. The skeleton is formed by a regular alternation of phosphoric acid and a deoxyribose molecule. Taking P as phosphoric acid and D as deoxyribose, the DNA skeleton looks like this:

$$\ldots - P - D - P - D - P - D - P - D - P - D - P - D - P - D \ldots$$

Each sugar molecule D is attached to a base (A, C, G, or T) in a sequence that changes and distinguishes that particular section of DNA.

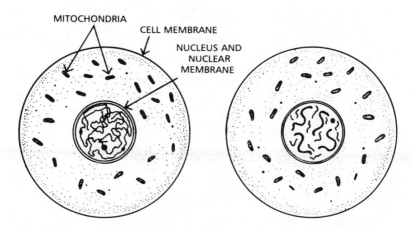

3.6 A simplified cell structure. *Left*: When at rest, the DNA in the cell's nucleus is a mass of strands. Mitochondria resemble bacteria and are found outside the nucleus. Each one contains one or more tiny strands of circular DNA (not shown). All cells have at least one mitochondrion, but there may even be tens of thousands. *Right*: When preparing for cell division, the strands of nuclear DNA that form the chromosomes become shorter and thicker, thus becoming visible under the light microscope.

For example, the full structure of a single piece of DNA may look like this:

$$\ldots -P-D-P-D-P-D-P-D-P-D-P-D-P-D\ldots$$
$$\quad\quad T \quad\quad A \quad\quad A \quad\quad C \quad\quad T \quad\quad G \quad\quad C$$

DNA can be divided into sections, each of which contains a D and P molecule plus one base (A, C, G, or T). These sections are called *nucleotides*, and there are four types because there are four bases. We will use this new term because it is a little less generic than *base* and has been widely adopted.

DNA can be thought of as a chain of nucleotides, a little like the string of pearls we used in Chapter 2 to explain proteins and the molecular clock, but DNA consists of four types of pearl instead of twenty.

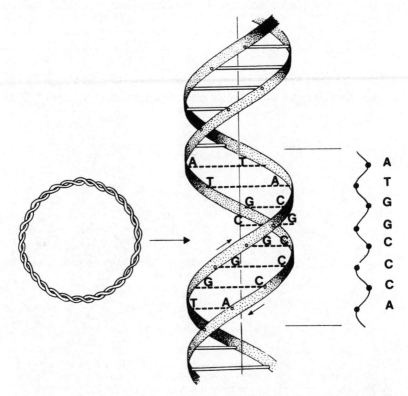

3.7 A round mitochondrial chromosome. The DNA it contains forms a double helix. Each helix is a strand made up of evenly spaced nucleotides. There are four types of nucleotide—A, T, C, and G—with a unique sequence. Each mitochondrial helix has about 16,500 nucleotides.

The sequence of DNA nucleotides is responsible for the *entire* biology of an individual. The order of nucleotides in the chromosome dictates everything from hair color and height to the size of the nose. Everything determined by heredity is registered in the chromosomes, which are made of DNA: consequently, all hereditary features are held in the DNA and depend on the sequence of the bases. DNA is like a book specifying the biological identity of each individual.

Mitochondrial DNA

The nucleus of the cell carrying the mitochondria contains a set of chromosomes that direct the duplication of cells and their function. The mitochondria are unable to grow and multiply on their own. A billion or more years ago, they were free bacteria that subsequently adapted to live in symbiosis with the cell: a cell cannot normally do without mitochondria, nor can the mitochondria do without the cell.

Each mitochondrion contains one or more strands of DNA, in the form of a tiny ring (another characteristic of bacteria, which generally have a single round chromosome). The tiny chromosome inside a human mitochondrion comprises about 16,500 nucleotides and is therefore much shorter than any of the chromosomes in the nucleus (which has tens or hundreds of millions). The mitochondrial chromosome serves to make other mitochondria, but only in association with the cell nucleus, which complements and supervises, as it were, the overall situation. This prevents the mitochondrion from going wild and becoming a sort of uncontrollable growth within the cell. Many of the mitochondrion's component substances are manufactured within the nucleus, and this ensures integration between nucleus and mitochondrion.

An important point to remember is that *the mitochondria are passed to children only by the mother*. Two children from the same mother will have identical mitochondria, even if the father is different. Every so often the mitochondrial DNA changes, or mutates, very slightly and one of the 16,500 nucleotides is replaced by another. From then on, all the descendants of that woman will receive the mutated strand in its new form.

Mutation is a fairly rare phenomenon. When we look at the mitochondrial DNA of people related on their mother's side, we notice no differences, but if we take individuals with no obvious blood relationship, the differences spring to our attention. A sequence of 16,500 nucleotides is too large to study in one shot. We can easily, however, make comparisons on the basis of a representative portion.

Looking for Eve

Allan Wilson and his students' best work involved a single, not very long, section of mitochondrial DNA, comprising six to seven hundred nucleotides. This is a limited sample, but varied enough to render the discovery of two identical individuals extremely unlikely, unless they are actually siblings or very closely related on the mother's side.

The differences observed can be used to draw a diagram or family tree, since individuals with only one different nucleotide have a more recent common ancestor than those with two or more different nucleotides. We have already seen this method used to study the evolution of proteins in the last chapter. It is also the basis of the so-called molecular clock.

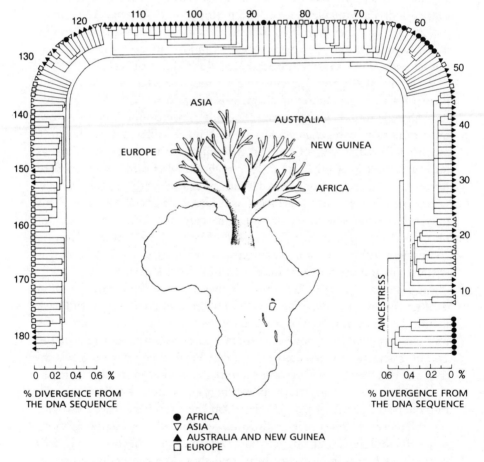

3.8 Allan Wilson and his team's tree reconstructing the origin of mitochondria in 182 individuals on the basis of the structure of their mitochondrial DNA.

Wilson's tree looks something like a horseshoe. The 182 people studied are arranged around the outer rim and divided according to origin. There are Africans, Europeans, Asians, and natives of Australia and New Guinea, and the tree is a result of close analysis of the sequence of nucleotides in the sampled section of mitochondrial DNA. The earliest person from whom all 182 descend is indicated at the bottom (ANCESTRESS). Where and when did she live? The tree can answer both these questions. The ANCESTRESS bottom right has two lines of descent. The lower one leads after several divisions to seven individuals, all of whom are African. The other branch heads upward and leads, after a multitude of branchings, to all the other members of the sample group, including many Africans. A quantitative idea of the diversities among individuals is provided by the scale, which gives the total number of differences as a percentage of the total number of nucleotides in the DNA sample. As with the proteins, the assumption is that the number of nucleotides differing in two individuals increases on average in relation to the temporal distance from their last common ancestor.

The first and oldest fork on the tree, which corresponds to the last common ancestress, separates Africans from other Africans. The branchings separating the people of other continents all come later, suggesting that modern humans originated in Africa.

This conclusion has attracted criticism because the same data could be used to draw up many different trees, some of which may be better and not point to Africa as modern human's first home. In fact, this reconstruction is not the safest method of dealing with the question. There are many methods of tree construction and many different potential results; however, nearly all the other methods and a number of other data point to the same conclusion.

Another, quite different analysis also concludes in favor of African origins. Among the peoples of various continents, Africans have been shown to be by far the most heterogeneous group. It is reasonable to expect that the oldest population will display the greatest diversity. This criterion operates independently of the method used to reconstruct the tree, which may be considered unsatisfactory for both theoretical and statistical reasons.

Allan Wilson and his team then calculated the date of birth of the so-called African Eve who is the common ancestress of all the 182 individuals (Eve rather than Adam because mitochondria are transmitted only by women). To do so, they compared the genetic difference between descendents from the first fork leading from the common ancestress to the first Africans and others to that observed between an average human and a chimpanzee. Wilson himself had already shown that the separation between these two distant cousins happened around five million years ago, and he calculated that the first separation in the human mitochondrial DNA tree took place 190,000 years ago. Because of the broad margins of error involved, in practice this date was taken as somewhere between 300,000 and 150,000 years ago.

Dating the Birth of Modern Humans

The idea is that humankind as we know it began in Africa. *Homo habilis* first appeared in Africa and so did *Homo erectus*. Archaeological observations suggest that modern humans, too, originated in Africa—or perhaps the Middle East—and later spread beyond. So Africa is a constant element. The question is when. For *habilis* we talk of two and a half million years ago, and *Homo erectus* left Africa at least a million years ago (some now say two million); for *Homo sapiens sapiens* (modern humans) we tend to use a date of one hundred thousand years ago or more, since we know that by that time modern humans resided in both Africa and Israel.

Allan Wilson's figure of 190,000 years is compatible with the latter date and with African origins. However, we have not yet explained the difference between the 190,000 years suggested for mitochondrial DNA and the age of one hundred thousand years attributed to the oldest modern human remains in South Africa and Israel.

Wilson's 190,000 years does not refer to the *first modern human*. This number actually corresponds to the time of the first mutation of mitochondrial DNA whose existence we can recognize, i.e. to the first fork line, in Allan Wilson's tree and thus to the common mitochondrial ancestress of everyone living today. Therefore, a very particular event took place 190,000 years ago; one egg of one woman happened to contain a mutation in mitochondria, something as simple as one of the 16,500 nucleotides being replaced by another. This egg generated a daughter who passed the mutated mitochondrion to her children. The destinies of that woman's descendants varied greatly in geographic terms, but it seems probable, although not certain, that she herself lived in Africa. Those of her descendants who left Africa did not necessarily do so immediately after that mutation. On the contrary, a considerable amount of time must have gone by before a group of her descendants started from some (unknown) African location, and reached the Middle East via the Suez isthmus, or Arabia via the Red Sea. The migration may have occurred not long before the owners of the first modern skulls found in Israel lived, and it is highly probable that this was a long time after the mitochondrial DNA mutation in the "ancestress" that heads Allan Wilson's tree. It is right for the two dates, mitochondrial and archaeological, to be different, and the former must precede the latter historically, possibly by a long interval.

African Eve: The Parallel Example of Surnames

We can understand African Eve better by applying the example of surnames. In Western culture, surnames usually are taken from the father —in contrast to mitochondria, which come from the mother. Many people have thought that Allan Wilson's study implies we all descend from one single woman. This is the first misunderstanding generated by the use of Eve, facilitating the reaction that "in that case, there was a single progenitress." The name has proved very popular with journalists because of its solemn and portentous overtones: according to the Bible, Eve was the first woman. None preceded her, and everyone descends from her. This may be true as far as mitochondria are concerned, because the mitochondria of only one woman have survived, whereas those of all her contemporaries have been lost. We have already said, too, that mitochondria are passed down the female side. Let us consider the example of surnames, which traditionally are also passed down on one side only (the male side), to further clarify the issue of African Eve, who is a "mitochondrial," not biblical, Eve.

The number of surnames in small mountain villages is often very limited (we will talk of this again later because it is a very important phenomenon in genetic evolution, as well known as genetic drift). If this elimination process were to continue long enough, eventually the whole village would have the same surname. This happens very rarely in Europe, because surnames are generally only a few centuries old, and the effects of this phenomenon are slower to emerge. In China, however, surnames are frequently very old indeed (some more than four thousand years in age), and there are villages and even small towns where everyone has the same surname. Were this to continue for many millennia, everyone in China would eventually have the same surname, and finally everyone in the whole world. Obviously, at that point surnames would cease to serve their purpose and would be abandoned as a system of identification.

In exactly the same way, all the mitochondria on the earth today descend from that carried by one woman, "mitochondrial Eve." It is as if there were only one surname, except that the mitochondria descending from Eve are no longer identical because they have mutated a little through the generations. These mutations are what allowed Allan Wilson to reconstruct his tree.

The great majority of our genetic makeup is contained within the chromosomes in the cell's nucleus, to which this principle cannot be applied because chromosomes are inherited from both the mother and the father (except for the X and Y chromosomes, which follow special inheritance rules and are responsible for sex determination). Mathematical analysis of our evolution shows that among the ancestors contributing to humans there have always been a very large number of men and of women, let us say between ten thousand and one hundred thousand—so no Adam and Eve in the biblical sense.

In 1992–1993, criticisms of Allan Wilson's tree and dating of the common ancestor appeared, casting doubt on the significance of the Berkeley team's conclusions, but the validity of such criticism has been considerably reduced by further work. Three separate sets of observations regarding different sections of mitochondrial DNA currently support the idea of African origins. Some laboratories think Wilson's 190,000 years should be greater, but very recent Japanese and American research point to later times. Both continue to show Africa as the modern humans' birthplace, and analysis of chromosomes in the cell nucleus also reinforces this idea.

Science and Certainty

The so-called person in the street often looks to science for certainty. Scientists, for their part, dedicate much of their energies to the opposite activity, by raising doubts and changing ideas. There are plenty of religions and ideologies willing and ready to lay claim to the "truth." To understand the process of science and how certain, or uncertain, it can be, laypeople need to know that there are two profoundly different types of scientific work: experimentation and observation.

Experimental research allows you to achieve the degree of certainty you desire. Conclusions already reached can be reinforced and refined by repeating the experiment on a greater scale and by modifying the conditions. The results may conflict with the original conclusions, necessitating a change of standpoint. In various disciplines, vast knowledge has now been accumulated. Without ceasing to experiment, certain conclusions have been reached which either never change or change so negligibly that they can be relied on as absolutely valid, after clarification and refinement.

In modern biology, DNA is the pivot of the explanation of the phenomena of life, including the origin of living organisms, and this reflects a vast number of observations and experiments, some of which are fairly indirect. It is now unthinkable to repeat *all* the experiments that have contributed to certain conclusions in order to be absolutely sure they are valid. If we have doubts, one extremely good way of persuading ourselves that a theory works is by applying it. To take an example from biology, theoretical knowledge about DNA has allowed us to insert the gene responsible for the production of a complex substance, such as an enzyme, hormone, or growth factor, into a bacterial chromosome. Many genetic engineering experiments of this kind have successfully made bacteria produce substances characteristic of the human organism that have the same structure and functions as those produced by humans. Without such experiments, it would be extremely difficult to obtain enough of certain substances for clinical use. This kind of application is the best proof that the theory behind it is correct. At the same time, it is reassuring to be able to land on the moon and obtain hard evidence to confirm certain conclusions: in fact, scientists were right to oppose earlier views that the moon is made of cheese or is the home of the dead.

In another type of science, however, the only possible approach is that of observation, since experimentation is impossible. This is history, which reconstructs facts that have occurred in the past. History cannot be repeated at will, nor can we run an experiment relaunching the evolution of the stars or living organisms or humanity, to see if the moon or the human species would be the same or a little different, a second time around.

We can run tests of evolution, but these are vastly simplified in relation to the history of which we are a part. We can take drosophila, the tiny fruit fly on wine must that is the organism most used in genetic studies. We can raise them in bottles or other controlled conditions, create populations of thousands, and watch them remain the same or change, lose their wings, or grow extra ones. We can use computers to simulate ever more complex evolutionary processes. We have a sound mathematical theory of evolution that allows us to make far-reaching predictions and check them. But for all our theories and simulations, we can never hope to imitate history as it happened because it is far too complex and detailed a phenomenon. Furthermore, chance plays a significant role in historic events, and our methods allow us to predict only average situations, not individual cases. As a result, historical interpretations are doomed to greater uncertainty than experimental knowledge. The hope of having and leaving no shadow of a doubt is much more fragile.

That there still are people who say that evolution never took place is not surprising, either. This view tends to be held by people with no scientific knowledge whose convictions derive from a quite different source: their religion, which teaches that evolution has not happened. Some faiths are sufficiently flexible not to exclude the idea, but others are bound hand and foot. For example, acceptance of evolution is simply incompatible with a literal interpretation of the Bible. Certainty about evolution is the prerogative of those who have objectively studied and considered the subject.

To be convincing with our historical interpretations, we must accumulate as much evidence as possible. However, there will always be those whose ingrained prejudices will lead them to refuse to accept certain concepts, even in the face of the strongest evidence to the contrary.

Historical sciences are therefore subject to a greater degree of uncertainty than other disciplines, and the only hope of overcoming this limitation is to observe each phenomenon from every possible standpoint. This is one reason I find it necessary to study human evolution not only from the point of view of biology but also in terms of archaeology, social anthropology, linguistics and so on. I will be the first to disbelieve my own ideas if a scientific discipline points in a different direction. The production of just one incontrovertible proof to the contrary will destroy a whole theory. If that proof exists, I want to find it, whatever its nature. The only way of controlling our scientific assumptions is to observe them afresh from new angles to see if they match the results expected, resting only if a theory effectively allows us to predict the result of all reliable observation. For this reason it is essential for our research to take into account all related fields.

It is undoubtedly true, too, that science is always accompanied by some degree of uncertainty. If a doubt is not raised sooner, it will be raised later, which explains why our statements are usually qualified by "probably" or "apparently" and so on: scientists know very well that even the tiniest alteration can demolish an interpretation, and they endeavor to avoid this by making their own changes in order to look at phenomena under different lights and highlight other possible interpretations. Scientists who experiment with the present or deduce from the past are historians in a broad sense. They are bound to work immersed in doubt; they formulate theories to explain phenomena, then test and check them as much as is humanly possible, and they have to be ready to change their position, or abandon it entirely if necessary. They are also more likely to have to reconsider their conclusions than experimental scientists.

Looking for Adam

The example of surnames has shown how we can use mitochondrial DNA to trace a special kind of Eve. Is there a similar way of tracing Adam biologically? Yes: by using the Y chromosome. This chromosome is found only in males, for the simple reason that it determines the male gender. As we have already said, DNA is the physical basis of hereditary makeup; it is contained in twenty-three pairs of chromosomes inside the cell's nucleus and contains billions of nucleotides (as against the 16,500 of the mitochondrial chromosome found between the nucleus and outer membrane). Each pair of chromosomes is made by one chromosome from the mother and one from the father. These are indistinguishable under a microscope, except for a pair known as XY, which is found in males. X and Y chromosomes are easily distinguishable because X is a medium-size chromosome, while Y is one of the smallest. The Y chromosome can be colored to tell it apart from the other small chromosomes in the nucleus. In females one finds two X chromosomes.

So the equations for gender are as follows:

$$\text{Male} = XY \qquad \text{Female} = XX$$

Spermatozoa are the long-tailed cells in semen that swim through the vagina and uterus toward the fallopian tubes looking for an egg cell to fertilize. When sperm is formed, X and Y are separated, and each spermatozoon receives either an X or a Y chromosome. The egg cells, however, which join a spermatozoon to create an embryo, can receive only an X chromosome. A spermatozoon with a Y chromosome that meets an X chromosome egg cell will make an XY embryo, a future boy, that is. An X chromosome spermatozoon will make an XX embryo, which is a future girl.

Like the mitochondrial DNA, the Y chromosome has also mutated, and there are differences between individuals. For this reason, we can hope one day to reconstruct a family tree for the Y chromosome and date the carrier of the first surviving mutation—an Adam, or Adam Y by analogy with the mitochondrial Eve. If we manage to pinpoint the Y-chromosomal Adam's date of birth, it will probably be different from Eve's. That would make it still more difficult to believe in the biblical couple who were expelled from the earthly paradise.

FOUR

Why Are We Different?
The Theory of Evolution

If we choose two people at random, there will always be some difference between them, no matter how small. How do we explain this? These are not differences in race, which spring immediately to the eye. The fact is that in any town or village in the world, we find that people always vary considerably.

Some of the differences will be due to accidental factors or voluntary changes. For example, nowadays there are people with green hair. This relatively recent phenomenon is attributable to a dye bottle, and not to any biological change. But it is easy enough to believe that other differences are truly biological, (genetic, that is, hereditary or innate). With varying degrees of meaning, these words signify that these differences are dictated by nature, or, using synonymous expressions, by our DNA, by chromosomes, or by genes.

Some striking similarities among relatives convey very forcefully evidence that biological inheritance can be extremely precise. Children sometimes resemble each other or one of their parents with respect to some traits to an astounding degree. Their hair may fall exactly the same way and they may share the same tics or habits that are otherwise very rare. Then there are identical twins. Those called "identical" are truly identical, and it is more than correct to say they are like peas in a pod. Greek and Latin dramatists, and Shakespeare himself, have all successfully exploited the confusions arising from the extreme resemblance of identical twins. They can be so alike that even their mothers or partners need to check special marks to tell them apart. They are a demonstration of the amazing power of biological heredity.

The mechanisms of heredity are responsible for the similarity between parents and children or twins. Likewise, these same mechanisms are responsible for the differences between individuals. It is the existence of differences between individuals that make us wonder why there are also extraordinary similarities in certain circumstances. This raises the question: how do parents transmit hereditary features to their children?

Germinal Cells: DNA, Genes, and Chromosomes

Every living organism is composed of cells, each of which has a particular function. A human being is made of a million billion cells. The creation of children is managed by special cells, known as germinal cells because they create the germ or new cell that forms an individual of the next generation; in a male this cell is the spermatozoon, in a female the egg cell. A new human is conceived when a spermatozoon fertilizes an egg cell. These germinal cells contain all the DNA that forms the new person, and they make him or her similar to the parents.

Spermatozoa are made of a nucleus containing the chromosomes, a protective membrane, a long tail to swim and to penetrate the egg cell, and little else. Egg cells are fairly large by comparison, but the nucleus containing the chromosomes is about the same size as the spermatozoon's. The hereditary features that children receive from their parents are encoded in the chromosomes, in a substance, DNA, that makes biological transmission possible. Biology in its entirety confirms this. It is not surprising, therefore, that the same DNA is found within every cell in an organism, not just in the germinal ones.

DNA can be thought of as an instruction manual for drawing up a new individual. The manual is "read," as it were, by the cell mechanisms that use information in the DNA and physically create the child by transforming the material available for the purpose (the nutrients provided by the cell). The DNA required to make a human can be imagined as an enormous encyclopedia in many volumes (one volume for each of the twenty-three chromosomes in each of the two germinal cells). Every volume is much longer than a normal encyclopedia because it contains a far greater number of words.

When looking at mitochondria, we saw that this book of DNA uses an alphabet of its own, made of the letters A, C, G, and T, the nucleotide types. A DNA strand is made up of a row of nucleotides whose order is unique to every individual.

This is the chemical basis of heredity. What we call a gene or hereditary factor is a segment of DNA that has a special function and that is generally made up of a few hundred nucleotides (although there may be many more) in a particular order. The same four nucleotides are used, but their sequence changes in every section and determines the function of each gene. One gene forms the pigment giving a certain hair, eye, or skin color; another is responsible for the formation of one or more of our body's numerous visible and invisible features; and so on.

Every germinal cell has a complete set of chromosomes, one for each type, so the spermatozoon has twenty-three chromosomes and so does the egg cell. Their fusion gives rise to a new individual whose cells will have forty-six chromosomes, or, to be more precise, twenty-three pairs,

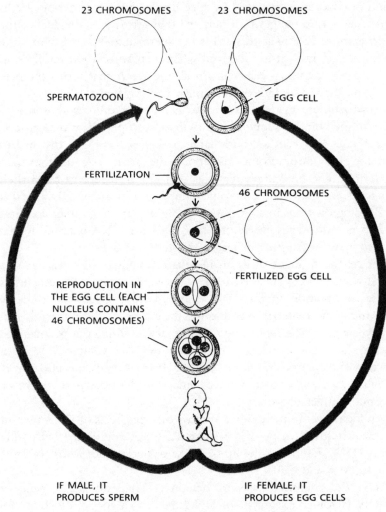

4.1 Fertilization of the egg cell and formation of an embryo.

since each chromosome has its own separate identity, and the new human receives one copy of each from both its father and its mother.

The Transmission of Genetic Features

Here is a brief glossary of the terms needed to understand hereditary transmission: *DNA* is a chemical substance with a precise structure; it is a strand made up of four nucleotides with a particular order, like the letters of words in a book. A *gene* is a segment of DNA that plays a special, recognizable role in the development and activities of a living organism; it is generally hundreds or thousands of nucleotides long, but sometime much more. A *chromosome* is a very long strand of DNA within the cell, wound in a unique way that makes it visible and recognizable in certain conditions.

A large number of genes are linearly arranged on the chromosome. There are also sections of DNA we do not call genes, because we do not know their function or even whether they have one. Some of them may be harmless parasites, sometimes called "selfish DNA."

Living organisms contain a number of chromosomes, ranging from one in bacteria to dozens and even hundreds in superior organisms. Human cells contain twenty-three pairs, as we have said: of each pair, one is provided by the mother and the other by the father.

All the DNA contained in an organism—and responsible for heredity—is held in the chromosomes, which are found within the *cell nuclei*. There is a little DNA outside the nucleus in the mitochondria. *Mitochondria* are small "organisms" with a short piece of DNA forming one tiny circular chromosome which are transmitted to both male and female children by the mother alone (see the question of African Eve in Chapter 3).

Chromosomes are found in every cell of a living organism, but only the chromosomes in the germ cells are transmitted to the next generation. Once fertilized, the egg cell splits into two cells, which split in turn and create four cells, which split into eight and so on, in a multiplication process that generates the new individual. The fertilized egg cell and all the body cells that derive from it usually have forty-six identical chromosomes. A small part of the cells formed in this way will eventually make either sperm or egg cells in the new individual.

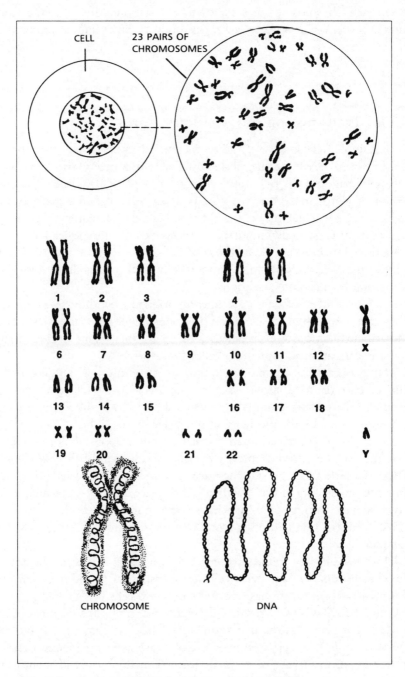

4.2 Cell and nucleus with detail of chromosomes and DNA.

Mutation

During the duplication process, the DNA of the first cell is copied. The two copies of DNA transmitted to the "offspring" cells are identical to the original DNA. However, mutations that are simply errors in the copying process can take place. These can occur in the division of any cell, including those that become sperm in a man and egg cells in a woman. These are the *germinal* cells that ensure the transfer of biological inheritance to progeny. We will examine the mutations in germinal cells, since they are the ones that can be passed on from one generation to the next.

We have already said that DNA is a sequence of a huge number of A, C, G, or T nucleotides. A mutation is the alteration of one of these. For example, a father may have a segment of DNA in a certain position with the following sequence of nucleotides: GCACCAATC. If there is a mutation in the third nucleotide in the sperm that will create a child, and a G replaces A, the new baby will have the following sequence: GCGCCAATC. Not only the new baby is affected, however. From that moment on, all the baby's descendants, if any, may also have and transmit the mutated DNA.

The Consequences of Mutation

In some cases, the tiniest alterations to DNA can have enormous consequences; they can determine the difference between health and sickness. (The same principle can be seen in other fields. The failure of one of the early space voyages reportedly was caused by one single misplaced comma in the launch program.) In other cases, there may be no consequence at all. It all depends on the position of the nucleotide and the nature of the change.

There are other examples from outside the world of high technology. The oracle at Delphi is said to have answered a king's inquiry about whether or not he would return alive from a certain military campaign with the following phrase, "Ibis redibis non morieris in bello." (The original is in Greek, but the ambiguity remains in Latin.) The Sybil's reply was verbal and therefore was not punctuated, and the king believed she said, "Thou shalt depart, thou shalt return, not in war shalt thou die." He departed a happy man and was promptly killed in battle. The answer should have been understood as "thou shalt depart, thou shalt return not, in war shalt thou die." Merely placing the comma after, rather than before, *not* changes the whole meaning.

A Genetic Disease

In a hereditary form of anemia, called thalassemia, the organism does not produce enough hemoglobin, the substance that gives blood its red color and permits the exchange of oxygen between lungs and tissue (we mentioned it when describing the molecular clock). The production of hemoglobin is controlled by two genes that form two proteins (or protein chains), known as alpha and beta.

The genetic defect responsible for a particular form of thalassemia common in the Mediterranean, and especially in Sardinia, affects the gene responsible for synthesizing the beta chain. The DNA of the beta chain is made of 438 nucleotides, and the defect concerns the 118th nucleotide, which normally is G. A mutation, which probably took place in Sardinia three or four thousand years ago, has replaced the G with an A. The result is that the beta chain gene is not "read" from that point on. Synthesis of the chain stops and normal hemoglobin cannot be produced.

Until recently, a hundred or so children a year died in Sardinia as a result of this mutation, which has in the meantime spread into a large number of the island's inhabitants.

In the Medieval Monasteries

A mutation is a change in DNA. Since DNA is copied from one generation to the next, when a change occurs the next copy starts with an altered version because it uses DNA that has undergone mutation. People who have a mutation, then, pass it on to their own children.

Virtually the same thing happens in literature, in the way that mistakes creep into manuscripts. On feeling he was about to die, the famous sixth-century Irish monk the Venerable Bede wrote a poem about what may happen after death, and asked if the approach of death allows us to understand it any better. (Not surprisingly, he concludes that the answer is no, not in the slightest.) Here are the opening words of the poem in ancient English, as they have come down to us in seven different manuscript copies. They all look very similar, but there are eloquent differences that allow us to reconstruct their history.

Manuscript	Century	Opening Words
1	9	FORE TH'E NEIDFAERAE
2	10	FORE THAE NEIDFAERAE
3	12	FORE TH-E NEIDFAERAE
4	12	FORE TH-E NEIDFAER-E
5	15	FORE TH-E NEYDFAER-E
6	13	FORE TH-E NEYDFAOR-E
7	12	FORE TH-E NEIDFAOR-E

"Before the inevitable journey"

The seven texts are arranged according to their degree of similarity and not chronologically. The hyphen shows where a letter has dropped out, so in manuscripts 3 to 7 the second word is "the," as in modern English.

In each case, a monk is copying from an earlier manuscript. Every so often the copier alters a spelling, either by mistake or because he prefers it that way. Each change has been picked up by subsequent scribes, who are now copying a manuscript different from the original.

The analysis works as follows: of the three opening words, the first, *fore* ("before," "afore"), is always written the same way and is therefore of no help to us. The second word is identical in manuscripts 3 to 7; the third—composed of two words, *neid* for inevitable (necessity or need) and *faerae* for journey—reveals a common relationship between 1 to 3 on the one hand, and 4 to 7 on the other. There is, therefore, some uncertainty regarding the position of manuscript 3, which resembles 1 and 2 in some respects, and 4 to 7 in others. This doubt can be resolved by studying the rest of this long poem.

A philologist has made the following reconstruction of the derivation of the manuscipts. A seventh-century manuscript, which has not come down to us, was copied by two monks. One of these copies was the basis for manuscripts 1 and 2 only, while the other became the basis for the remaining five copies: first numbers 3, 4, and 7; then number 6, which was copied from 7; and number 5, which was copied from 4. This analysis uses the same reasoning as that applied to reconstructing molecular evolution and described in the examples of hemoglobin and mitochondrial DNA.

An error often has no consequence at all, as in the case of this poem. In fact, the meaning of the poem by the Venerable Bede's is unaltered by the changes.

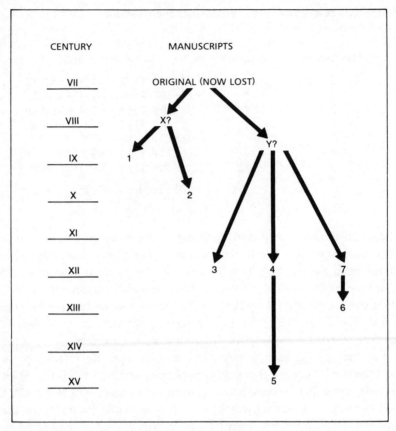

4.3 Derivation and order of seven medieval manuscripts.

An Unexpected Blessing

Sometimes a casual variation such as a mutation can bring advantages, just as a wrong turn sometimes ends up being a shortcut to our destination. The same with mutations: sometimes, albeit rarely, they are beneficial.

This is to some extent the case with thalassemia (Mediterranean anemia). A serious anemic disorder, it used to be fatal in all cases. Even today, repeated blood transfusions (which carry with them the associated risk of contracting hepatitis B or AIDS) are needed for a sufferer to survive. Thalassemia is found in certain parts of Sardinia and other coastal areas of the Mediterranean and beyond, and it is natural to ask why it continues to exist, if it is fatal. Surely a hereditary genetic disorder should disappear of its own accord if it automatically impedes reproduction by killing those affected—shouldn't it?

There is an explanation. For a child to suffer from thalassemia, both the mother *and* the father must be not quite normal genetically. Both, in fact, must carry the thalassemia gene, although they usually show no outward signs of it. People of this kind are known as healthy carriers and have inherited the gene from only one of their parents. As a result, not only do they avoid the anemic disorder, but the presence of the thalassemic gene in their makeup actually gives them stronger resistance to malaria.

Malaria was once an extremely serious illness found in various parts of the Mediterranean, including Sardinia, particularly on the coast. Healthy carriers have a great advantage in malarial regions. That is why the mutated gene has spread to the descendants of the first carrier. The anemia itself affects only those who inherit the gene from *both* parents, which can occur, of course, only if both parents are healthy carriers. Healthy carriers are not common, so marriages between them are not common either. As a result, the disease itself is relatively rare, generally less than 1 percent of the population, and only in areas where the risk of malaria is high. The percentage of healthy carriers can reach 20 percent.

Probability states that the condition will afflict one child out of four in a couple of healthy carriers, and in the past that child always died in early childhood. Another two children will be healthy carriers of the gene (and therefore more resistant to malaria). The fourth child will be perfectly normal (and therefore lack the protection afforded against malaria).

High incidences of thalassemia always coincide with areas where there has been a high risk of malaria in the past. The loss of the children born with thalassemia is more than compensated in unkind, but very concrete, demographic terms by the benefits to the parents in terms of survival in malarial regions.

The Human Genome

Our chromosomes contain about three billion nucleotides in all. Some, such as certain amphibians, have more DNA than we do, but this doesn't make them more intelligent or more complex. Some of the DNA may be useless, and is variously called "junk" or "selfish" DNA, and so on. It is difficult to tell exactly which DNA is useless, or actively harmful, but we do know of cases where this selfish DNA causes illness. Insects have less

DNA than humans, bacteria less than insects, and certain viruses least of all. Viruses are parasites that—like all parasites—have been simplified during the course of evolution until the only function of which they are capable is that of finding and sticking to a host organism in order to reproduce. They have often developed extremely complex and efficient mechanisms for entering an organism. They do not need many other biochemical mechanisms to produce food, because they exploit those of the host. They only need to enter a suitable host, multiply in it and go out again to look for another victim.

A mutation almost always takes the form of the replacement of one nucleotide by another. Occasionally, though, one or more nucleotide may be lost or one or more may be added.

DNA is shaped in the famous double helix, which most people now recognize. Figure 3.7 showed its exact appearance (there is a monument to DNA in the form of this double helix in a Beijing city square). The double helix is composed of two spiraling strands of DNA that are perfectly complementary. For example, if there is a C element in a determined position on one spiral, there will be a G on the other, and vice versa. A and T also pair in this manner. The two strands forming the double helix are therefore completely different, but the information contained in one strand corresponds exactly to the other in terms of the sequence of nucleotides. In substance the nucleotide sequences say the same thing and can be translated from one strand to the other without mistakes. For this reason, they are known as complementary.

This means that human DNA is actually made up of three billion *pairs* of nucleotides, found in the double helix that forms the twenty-three chromosomes in the germinal cells. We know the latter have only one chromosome for each of the twenty-three pairs; in genetic jargon, they have a haploid genome, which represents, as it were, the complete "samples book" of all the genes available in the human specie's hereditary makeup, a book that usually includes every gene in only one of the various forms in which it can exist in the different people in a population.

Since two germinal cells are needed to make an individual, in a person's cells you find both the genome provided by the mother and that provided by the father, meaning that there are two DNA double helixes (and therefore four nucleotides in each position), one from the mother and one from the father, making a total of twelve billion nucleotides.

We refer to three billion nucleotides, because in one way or another the other nine billion repeat the first three. The maternal and paternal contributions, however, are similar but not quite identical.

Are Mutations Frequent?

Normally, mutation is a rare phenomenon. It is rare because it has to be. We are built in such a detailed and complex fashion that a mutation can, and generally does, damage the organism because it is a random alteration. The possible consequences of mutation have been humorously compared to banging a faulty television with your fist. Sometimes it makes no difference, sometimes it improves the situation, and sometimes it makes matters worse. For a new individual to work properly, each constituent part must work properly. That is why it is important for mutations to be rare.

Mutation rates are affected by a number of factors, physical and chemical. Radiation increases them, the more radiation, the greater the dose that reaches the gonads, where gametes are formed (ovaries and testes). Test tube experiments have shown that many chemical substances can increase mutation rates. Whether these chemical mutagens, introduced with foods or as drugs, etc., are effective depends on whether they reach the gonads. Temperature also affects mutation rates—an increase causing an increase. Mammals and birds have constant body temperatures, and climate should therefore have no effect. But it has been suggested that the use of pants by men has increased the male mutation rate, by keeping testes at a higher temperature.

There are checking and correction mechanisms for the DNA copy made, just as a computer can check whether a file has been copied properly. DNA copying errors can be corrected with the result that mutations are infrequent. In a cell, in the course of a generation a few dozen nucleotides will mutate out of the total of three billion; this is a tiny incidence of error in the order of about one in two hundred milli n nucleotides at each copy.

Identical twins share exactly the same DNA because they come from a single egg cell that is fertilized by a single spermatozoon but splits into two *before* each half becomes an embryo. Identical twins have only a few dozen differences between them, which derive from mutations.

If we compare two individuals taken at random from a population, we will find a much greater number of differences, differences that have accumulated over generations. Naturally there are fewer differences between siblings or parents and children, although the level remains high, because there are always two genetically different parents contributing to a child's genetic makeup. It is interesting to note that the difference between siblings (or parents and children) is on average about half that between two people chosen at random from the same population.

Siblings can therefore be extraordinarily alike, but also very different, and this should not surprise us.

The average probability of one out of two hundred million nucleotide new mutations per generation is an approximate figure that is not yet accurately known. Two individuals taken at random from the same species have an average of roughly one in a thousand different nucleotides, which may not seem like many but corresponds to three million differences out of the three billion nucleotides in a human. If we compare individuals from different species, we find a greater number of differences, and the figure increases with the evolutionary distance between the species. When talking of proteins, we saw that a comparison of humans, horses, and chickens shows that humans are more like horses than chickens, and this is the basis for the molecular clock. The same concepts apply to the accumulation of differences in DNA and provide another molecular clock. We would expect it to, and it does, achieve the same results.

The Mutation's Destiny in the Balance

The mutation leading to the Sardinian thalassemia occurred in the gene responsible for synthesizing hemoglobin. We believe it to be a rare mutation. There are hundreds of different thalassemias, all of which have their own geographic distribution. This suggests that many of them derive from a single mutation that is different in each case.

Let us suppose that a certain mutation occurs for the first and perhaps only time in a spermatozoon. The egg cell it fertilizes has a normal gene in the position corresponding to the man's mutated one. The result will be an individual who is different from all the others previously existing in the population, heterogenous in terms of the gene in question because he or she has received a mutated gene (which we will call T for thalassemia) from the father and a normal one (N) from the mother.

Let us add two more new terms: *heterozygote* and *homozygote*. They are Greek composite words, *Zygote* meaning "fertilized egg cell," *hetero* meaning "different," and *homo* meaning "the same." *Heterozygote* describes an individual who has received a different form of a given gene from the father and from the mother; *homozygote* describes a person who has received the same gene from both mother and father.

Using N and T as symbols, the heterozygote is NT; there are two sorts of homozygote, NN and TT. TT is someone who has received the mutated T gene from both parents. This becomes possible only when there is a certain level of NT people of a reproductive age in the population. It takes some generations from the first mutation for a sufficient number of these to accumulate. Further explanation will clarify why.

First, however, we must pause awhile to introduce the idea of "fitness" in genetic types as applicable to the three groups NN, NT, and TT. Fitness describes the ability to survive into adulthood and procreate.

The word *fitness* has more than one meaning, so to avoid confusion we had best specify that we are talking about *Darwinian fitness*, because we are using it in the sense adopted by Charles Darwin, inventor of the theory of evolution through natural selection. In its more common usage, *fitness* refers to good physical shape achieved through the now popular activities of dieting, exercise, and so on. There is, of course, a connection between physical and Darwinian fitness, but while joggers measure their fitness in terms of the length or speed of their daily run, Darwin's fitness refers to genetic performance and is assessed in terms of the *average* number of children produced by a genetic type. It measures the ability both to reach adulthood and to have children. There are puny types who would cut a very poor figure in a gym but have lots of children, and there are splendid athletes who have none at all. Physical fitness and Darwinian fitness, then, are not the same thing.

The fitness of the three genetic types, NN, NT, and TT, changes depending on whether malaria exists in the region. If there is no malaria, the normal NN and heterozygote NT are equally fit in Darwinian terms, while TT has Mediterranean anemia (another name for thalassemia) and will die before reaching adulthood, unless he or she receives the very latest medical therapy. Recent and very costly treatment (such as bone marrow transplant) now gives sufferers improved life expectancy, and even the chance of surviving to the age of reproduction. Until not long ago, however, the chances of living more than a few years were very slim indeed.

The situation changes, however, if there is malaria in the area, because NT then survives better than NN. The malaria parasite multiplies in the red blood corpuscles, which it destroys before moving into others and destroying them in turn. This is the mechanism that provokes recurrent attacks of malaria. An NT's red corpuscles are different because of the mutation, and this makes it more difficult for the parasite to multiply, to the lasting benefit of the carrier's health. The TT type dies in any case.

It is clear, at this point, that the T mutation, while wreaking great harm on homozygotes, can have adaptational advantages, but only in the presence of malaria.

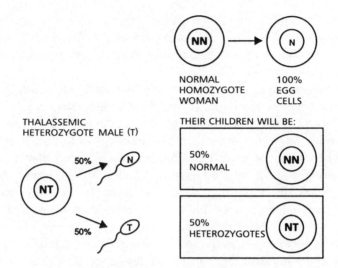

4.4 Combination of a homozygote and heterozygote, showing the children likely to be born from a marriage between a normal NN female and an NT male (or healthy carrier of thalassemia): half of the children are NN and the other half NT. These two genetic types are known respectively as homozygote and heterozygote. The outcome is the same for a marriage between an NN male and an NT (healthy carrier) female.

The Possible Destinies of Mutants

This lengthy introduction is necessary, but not sufficient, to understand what happens to the T mutation as soon as it is produced. Let us go back to the time when everyone was NN, except for one NT, who was every bit as healthy as the NNs and actually better off in malarial areas.

The first NT will perforce marry an NN, because there are no other genetic types available. He or she will transmit to each child *either* the N gene *or* the T. It is fairly clear that the two events have the same probability, 50 percent, exactly like heads and tails in the toss of a coin. The other parent is NN and can contribute only an N. As a result, on average, the children of an NN × NT marriage will be half NN and half NT.

Here we must congratulate readers who have managed to keep up with us because they have learned one of the fundamental laws of genetics, established by Gregory Mendel, abbot of the Moravian convent of Brno and published in 1865.

In practice, when one parent is NT and the other NN, half of the children are like one parent and the other half are like the other. It reminds me of Trilussa's humorous remark about the meaning of statistics: if I have eaten a chicken and you have not, then we have eaten half a chicken each.

If the NT individuals descending from the first mutant continue to increase in number over the subsequent generations—which is more than likely in malarial zones, because they have considerably higher resistance than NNs—at a certain point two NT types will marry and start a family. This introduces a new genetic possibility because a T spermatozoon can now fertilize a T egg cell. The result will be a genetic type hitherto unknown in the population, TT, who will fall sick and die.

Using the terminology we have learned, we can say that the marriage of two NT heterozygotes can lead to one TT homozygote out of every four children born (on average).

The NT × NT marriage, therefore, gives us one quarter TTs. A husband and wife who are both NT have the same chance of having a TT child as that of getting two tails from tossing two coins. The

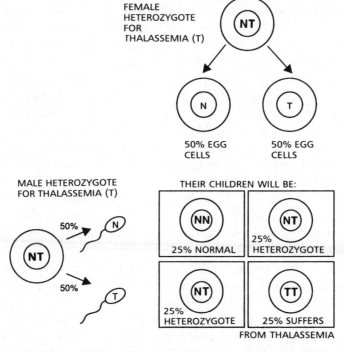

4.5 The combination of two heterozygotes, showing the children likely to be born from a marriage between two NTs (heterozygotes, or healthy carriers). On average, of their children one out of four will be normal, two will be healthy carriers, and a fourth will suffer from thalassemia.

husband must contribute a T spermatozoon (50 percent probability) and the wife a T egg cell (50 percent probability). This takes place in half of half of all the possible cases, or for one child out of four.

At this point, we have introduced another of Mendel's laws, and unfortunately we have also had our first case of thalassemia, which will soon be followed by others. About 110 patients were born each year in Sardinia, until recently. Nowadays, thalassemia can almost always be predicted and detected in time for termination of the pregnancy. The result is that very few thalassemics are born in Sardinia today. For those who are, it is usually because they were not tested before birth. They had either refused for religious reasons, or were born out of wedlock and parents could not be tested in time.

A Pause to Gather Our Thoughts

If you have understood and inwardly digested all the concepts presented so far, skip this section. If you are feeling a bit overwhelmed, however, the following summary may help you straighten your ideas. We have considered two different types of the same gene that helps form hemoglobin: N, the normal gene; and T, the mutated gene (thalassemic).

In people themselves, there are three combinations of these types: NN, a normal homozygote, who receives an N gene from both parents; NT, a heterozygote who receives an N gene from the father and a T gene from the mother, or vice versa; and TT, a thalassemic homozygote, who suffers from a serious disorder because he or she has received a T gene from both parents.

How fit are they? It depends on the environment. If the area is malaria-free, NN and NT are equally fit. TT is extremely sick and will not reach the age of reproduction unless he or she receives very special modern care. If there is malaria in the neighborhood, NN has a lower resistance to the most serious form caused by the most lethal parasite (*Plasmodium falciparum*) and is more likely to die than NT; as before, TT will die unless treated.

To return to the question of transmission, our heterozygote NT married to a normal NN is equally likely to have NT or NN children, because half his or her germinal cells receive the N gene and the other half the T gene, while the partner can provide only N genes.

Many generations are needed, and many NT descendants must derive from the first heterozygote before the first thalassemic is born.

In the initial generations after the first mutation, marriages between NTs are unlikely, because they are too closely related by blood. An NT can have an NT son and an NT daughter, but unions between brothers and sisters have been in disuse since the time of the pharaohs and the emperors of ancient Persia. After two generations, there could be a marriage between two first cousins, both of which are NT. The Catholic church, for example, allows first cousins to marry, although special authorization is needed. In reality, such marriages are not so rare (in many European countries and in some American communities, for example, the incidence ranges between one in a hundred and one in a thousand). After a few generations, the degree of blood relationship is forgotten. What happens when one NT marries another and they start a family? The parents both have one half T germinal cells; half of half of the children (25 percent, or one out of four) will receive the T gene from both father and mother, and therefore will have thalassemia.

A Mutation's Destiny: The Grand Finale

Let us return to the mutation that has caused the birth of an NT individual. How many children will this individual have? The range, obviously, is wide. Many people never marry, or marry but have no children, and in both cases the mutation disappears. If NT has only one child, there is a 50 percent chance he or she will be NN, which also would make the mutation disappear. If the child is NT, however, the mutation might possibly continue to be passed on. If the first NT carrier has several children, a number of whom are NT, the probability increases. Chance plays a very important part, especially so long as there are only a few NTs. After many generations, the incidence of the T gene may increase and with time reach a fairly high level, particularly if the population is a small one.

Without making complex calculations, we can easily see that in malarial regions (where NT has an advantage over NN, who succumbs to the disease), NT is much more likely to increase over time. In certain cases, the mutated gene may eventually supplant and wipe out the gene that was originally "normal," but calculations show that this takes many generations and it is unlikely that the N gene will disappear entirely, unless the advantage enjoyed by NT is really enormous. In the case of thalassemia, it is impossible for the T gene to supplant the N gene, because TTs die before reaching maturity. Even if many NTs are born initially, after a while the increase will end whenever there are too many TT children in the population.

The Forces Making Us Different

We can now start to answer the question posed at the beginning of this chapter: What makes individuals and populations biologically different? There are three factors: *mutation*, *natural selection*, and *chance*.

Mutations can be grouped into three categories: harmful (because they alter a function negatively), neutral (because they have no effect), and beneficial (because they improve the organism's ability to function in its specific environment). The first and the latter are examples of natural selection. A harmful mutation leads to the inability or reduced probability of the carrier living and reproducing normally. The exact opposite is true for the beneficial mutation. It increases survival or fertility, or both. The incidence of carriers of a beneficial mutation automatically increases over a number of generations.

The thalassemia example is complex because the presence of the T gene is beneficial where there is malaria, but only for the NT type. The TT variation is harmful and destroys its carrier. After a number of generations, a stable relationship between N and T will be established and a fixed proportion of TT children will be born, develop Mediterranean anemia (another name for thalassemia), and die. In this case, natural selection preserves a difference between individuals. In the absence of malaria, NN and NT are equally fit and natural selection exerts pressure on neither: it is an example of a mutation that is irrelevant or selectively neutral, at least as regards these two genetic types. Because TTs are unable to survive, the T gene is at a disadvantage and gradually diminishes, unless malaria comes to its aid.

Hereditary Diseases

A harmful mutation impedes normal development of an individual. Large mutations, such as those deriving from the loss of a chromosome or the production of an extra one, are generally fatal, sometimes even prior to birth, or lead to serious disorders, such as mongolism, which is caused by the presence of a particular extra chromosome. Even very small mutations that alter or add one nucleotide can prove fatal.

Hereditary diseases are caused by harmful mutations. An affected child is born alive but sooner or later falls seriously ill and generally has shorter-than-average life expectancy. Several thousand such diseases have been recognized, almost all of which are rare because their harmful nature tends to lead to their gradual elimination by selection.

Naturally, new mutations appear over time, and these mean there are always sufferers of genetic disorders in a population. Because mutations are rare, however, their number generally does not reach high proportions. Many genetic disorders are so rare that they have so far been analyzed in only a very few families.

At the other end of the scale, the most common mutation is that responsible for Down syndrome. This condition, which used to be called mongolism, leads to severe mental handicap, making it difficult for sufferers to be integrated into society. Down syndrome results when three copies of chromosome twenty-one, instead of the usual two, are inherited. This happens when a spermatozoon, or more frequently an egg cell, has two copies of twenty-one because of an error in the egg cell formation process. The syndrome affects one baby in a thousand, and more if the mother is over thirty-five. The incidence increases with the mother's age. It is possible to diagnose the syndrome in time to allow termination of a pregnancy, if desired.

A rarer, but still frequent, genetic mutation (the incidence is about one in three thousand births) causes a benign tumor, or neurofibromatosis, leading to polyps, generally on the epiderm. These can be numerous and reach considerable dimensions if not removed (one famous case of a nose that resembled an elephant's trunk inspired David Lynch's film *The Elephant Man*).

Other serious diseases—such as Huntington's chorea, which leads to progressive dementia and loss of coordinated movement—are rarer (one in twelve thousand births). Natural selection plays little part here, because the onset of the disease occurs fairly late in life, at about age forty on average, and leads to death around fifty. Its effect on fitness is virtually zero, because most children will normally have been born to patients well before the disorder manifests itself.

A large number of mutations are irrelevant, or neutral, in terms of natural selection. We do not even realize we have them.

Beneficial Mutations

Relatively few mutations are beneficial. The majority of useful ones that occurred in the past have been accepted by natural selection and have therefore become a part of us. If a mutation in the past gave greater resistance to a certain commonplace disease, sufferers from the condition died and only the resistant survived.

A significant example of a beneficial mutation in humankind's recent evolution is the ability of adults to use lactose. Lactose, the sugar present in milk, is an important nutritional element of human milk and that of all mammals, with a few exceptions such as the seal, whose milk contains no lactose. Lactose can be digested thanks to the presence of an enzyme—lactase—which breaks it down by separating out its sugar components. After weaning, all other mammals cease to produce lactase because it is no longer necessary. Many human babies continue to produce it. This allows them to use lactose and means that they can continue to drink milk into adulthood. A number of adults don't produce lactase, however, and as a result cannot drink milk without unpleasant side effects such as nausea, bloating, flatulence, and even diarrhea. Generally these people instinctively dislike and avoid milk, but unintentional consuming of foods containing it can cause attacks.

Until a few years ago, this milk intolerance, or more correctly, lactose intolerance, was not recognized or diagnosable. Sufferers are often unaware of their problem, at least in Europe and North America, where milk consumption is high: they can perhaps drink a little milk without being unwell or simply don't link the stomach upset to milk consumption. In areas such as China and many other parts of the world, milk is not considered a suitable food for adults and no one drinks it after the first years of life. Not all milk products necessarily cause a problem. The bacteria used in the production of cheese and yogurt consume most of the lactose present and the residue is generally low enough not to bring on side effects. Many dairy products are fine, therefore, even for those who have a milk intolerance. who have a milk intolerance.

History provides an explanation of this phenomenon. The consumption of fresh milk by adults has been possible only in the last ten thousand years, since the start of goat, sheep, and cattle farming. Prior to that, the only milk available was baby milk, which was given only to babies.

The raising of livestock does not automatically lead to drinking fresh milk, especially in adults. The development of certain customs has been necessary, and this has happened only in certain peoples of Europe (mainly in the north) and among some groups of African herders. The ability to produce lactase after weaning is a hereditary feature, the result of a mutation that was favored by the raising of animals and by the custom of people of all ages consuming the milk, without first turning it into cheese or yogurt.

With few exceptions, there is a very precise correspondence between the frequency of adults able to use lactose and the custom of drinking milk. In Scandinavia, where milk consumption is high, the incidence is 90 percent or more; in Italy it varies between 50 and 90 percent, depending on the region. In the U.S. the proportion of lactose-tolerant individuals varies with the ethnic origin of individuals. The ability to use lactose probably has several advantages in addition to that gained from the assimilation of lactose itself. For example, it helps the body absorb calcium, which is important in northern regions where the limited availability of sunlight increases the risk of rickets. (This once very common and serious condition has now declined enormously, thanks mainly to the wide availability and use of the antirickets vitamin, D).

Mutations Propose but Selection Decides

Mutation is casual and generates innovations that can be either beneficial or harmful. Selection decides the question automatically by promoting beneficial mutations and eliminating harmful ones, as determined by environmental conditions: mutations that confer advantages in the Arctic Circle do not necessarily do so in the tropics. The mutation that allowed lactase to continue to exist into adulthood is useless and, who knows, could even be slightly harmful if milk is not consumed after weaning, but it is beneficial if milk is a common food for adults as well as children. In this case, the environment is determined by a nutritional custom, which is a cultural and not a genetic factor.

There are other examples in which nutrition decides whether a type of gene is beneficial or harmful. It seems that a gene type common among Native Americans—and perhaps elsewhere, too—allows certain food components, particularly starch and sugars, to be stored in the body when food is scarce (just as camels and cacti store up water in the desert). When sugars or alcohol are plentiful, this type becomes prone to obesity and diabetes (or both), which explains why these two diseases are common among some Native American groups today. Presumably they were once unknown. Genes that favor certain types of storage are potentially harmful when storage is no longer necessary.

In Europe the development of agriculture has lead to the spread of cereals as the primary foodstuff over the last ten thousand years. Unlike meat, particularly fish liver, cereals contain no vitamin D. They do, however, contain a precursor that becomes vitamin D if exposed to the ultraviolet light from the sun's rays absorbed through the skin.

Cereal eaters can produce enough vitamin D to survive and grow normally if they are fair-skinned. Therefore, people can inhabit northerly regions where there is less sunlight, and continue to eat cereal products because a fairer skin color has been selected during evolution. In certain northerly regions, however, and even in the far north, some peoples, such as the Eskimos, have always eaten enough vitamin D from fish or meat to make fair skin unnecessary. On the other hand, if the skin is dark, the ultraviolet rays cannot get through. Dark skin affords considerable protection against strong ultraviolet rays, which could otherwise be harmful, but it stops the transformation of precursors into vitamin D. Where the solar intensity is high, however, enough vitamin D is still produced even if the skin is dark.

Another type of selection, which is less important in terms of strict survival but may be powerful is sexual selection. This is selection resulting from the choice of mates, and is not easy to study. It is influenced by tastes, which are individually variable, subject to fashion, and may change and become relatively unpredictable. Eye color may have been selected on the basis of taste, and the same may be true of hair color, which generally is linked to eye color (fair hair, light eyes; dark hair, dark eyes). As we have seen, there can be important adaptational components relating to climate and food in skin color, but also fashion and prejudice.

Factors of sexual selection may have caused skin variations. Perhaps rare types, be they fair or dark, have always been sought after. Rarity is often attractive, and mutation will ensure a wide range of the unusual. We don't know the original color of human skin. A legend told in China and elsewhere says that God created humans and baked them in an oven. His first attempt burned, and thus the Africans were created; the second attempt was underdone, resulting in the Caucasians; the third attempt was just right, and the Chinese were drawn from the oven.

Evolutionary Advantages

Natural selection guarantees survival of the fittest, the most suitable, that is, in one place and for one set of conditions; climate, nutrition, and resistance to disease are the most important factors. As we have already seen, it is better to be dark-skinned in tropical zones and fair in the north unless other rich sources of vitamin D are available. This is one reason for our different skin colors. Another is quite certainly resistance to sunlight; dark skin protects against heat rash and skin tumors caused by ultraviolet rays.

In their settlement of the planet, humans have gradually diversified to meet specific environmental conditions. The spread north and adaptation to survival in the cold occurred late in the history of evolution, and it was modern humans who made the innovations needed to penetrate the far north and arctic regions. Some of these innovations were biological, particular mutations, that is, introduced by chance and favored by selection. Various bodily and facial features form part of the most suitable biological design for certain climates. We have already seen how the pygmies are marvelously well adapted to the heat and humidity of the tropical forest.

At the other end of the scale—in the cold—it is useful to have small nostrils, because the reduced air flow gives cold air time to warm up before reaching the lungs. Extra eye protection is gained by making the eye slit smaller and surrounding it with pads of fat. Because heat is lost through the surface, the body's shape becomes shorter and wider (as ball-like as possible) in order to reduce the ratio of surface area to body volume.

The force of these biological changes has combined with important cultural developments: the use of fire, furs and clothing (skins sewn together with needles), warm huts, and a hundred other details, such as preserving food to survive through periods of shortage or covering the body with grease when the weather is coldest. Natural selection plays its part, since it has favored human types capable of making adequate cultural progress, but in this case it operates indirectly. However, to understand evolution fully, the role of both cultural as well as biological adaptation must be taken into account.

The Importance of Chance

There is a third, extremely important component of evolution. Its technical name is *genetic drift*, but it could easily be known as *chance*.

We have already said that a carrier of a mutation (or *mutant*) may die without children or have only normal children, in which case the mutation is lost. This can happen at various points in time. In fact, if we look at mutations several generations after their appearance, we see that only very few survive. It is a question of chance, and the opposite also can occur. One or more casual events can lead to a mutation spreading widely through subsequent generations of a given population and even actually supplanting the previous type.

Chance is especially important in small populations. To use the example of a tossed coin, with one coin there is a fifty-fifty chance of throwing either heads or tails, and two coins can very easily both land on heads. The probability, however, that ten coins all land heads up is about one in a thousand (or, to be more precise, half times a half times a half, and so on up to ten multiplications). If a hundred coins are tossed, it is most unlikely that all will turn up heads.

With a coin, the probability of heads and tails is the same at every toss. With genetic drift, the probability changes with every generation. If there is one mutant in a thousand people, it is not certain that we will find others in the next generation, since none may be born. For number lovers, a formula shows a 37 percent probability of a mutation being lost this way. The probability of there being only one mutant is also 37 percent; of there being two, 18 percent; of there being three, 6 percent; and so on. If there are three mutants in the second generation, the probability of the mutation being lost in the next generation is much smaller than if there is only one. If the frequency of mutants happens to increase, the importance of genetic drift will decrease.

The islands in the Pacific are usually inhabited by small numbers of people. Even the largest islands were originally colonized by small groups that subsequently reached larger proportions (such as the tens or hundreds of thousands of New Zealanders or Hawaiians). A classic case is that of the famous mutineers of the *Bounty*: six British sailors together with a similar number of Melanesian or Polynesian women (or a mixture of the two) "founded" the island of Pitcairn in the Pacific. After a stormy beginning in which almost all the founders quarreled and killed one another—after procreating—the population began to increase and migrate to other islands.

Demographic bottlenecks (drastic declines in population) were frequent on Pacific islands at various times, generally as a result of enemy invasion or typhoons. When there are only a few founders (or refounders) after a bottleneck, whose genetic composition may be different from that of the earlier people, significant changes in the genetic composition of a population can take place. The term for this phenomenon is "founders' effect," but it is still a special example of genetic drift. In reality, every generation is the founder of the ones that come after, and the effects of genetic drift accumulate through the generations.

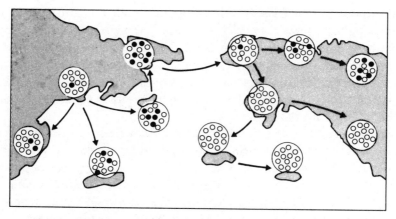

4.6 Change in gene frequency from one generation to the next as a result of genetic drift. Arrows indicate history of a colonization process.

Genetic drift can have dramatic effects on disease: rare genetic disorders may not exist in a founder group and will therefore remain nonexistent down the generations, unless they are reintroduced by a new mutation (which is unlikely) or by an immigrant (which is more probable). Alternatively, if a founder group does contain a mutation, the related disease can become extremely common. About 10 percent of the population of the island of Pingelap in Micronesia suffers from a serious eye defect causing total color blindness. This condition is almost nonexistent in the rest of the world. The mutation must have occurred in a member of the population several centuries ago and become common as a result of genetic drift. If there are ten founders and one has a rare genetic disorder that escapes adverse natural selection, because, like Huntington's chorea, it manifests late in life, the initial frequency of the disorder is 10 percent—which is extremely high. The frequency may then remain very high in subsequent generations.

Obviously, this principle is valid not only for geographically isolated peoples but for any sort of genetically isolated group. For example, if a community has few or no marriages with outside groups for historical, cultural, or religious reasons, it is quite likely to show an unusual pattern of genetic diseases: illnesses rare or absent elsewhere may be common, or vice versa. A very serious disease among Jews, known as Tay-Sachs disease, develops during the first months of life and is rapidly fatal. A case in a Christian family in South Africa some time ago took everyone by surprise: it was subsequently found that both the parents' families were originally Jewish and had only recently converted to Christianity. Like many other genetic disorders, Tay-Sachs disease can now be avoided through genetic screening, and embryonic testing if both parents are heterozygotes. Screening programs have been introduced for northern European Jews (Ashkenazi Jews), but not for other ethnic groups because it is so rare elsewhere.

In mountainous regions, where communities are small and migratory exchanges less frequent than in lowland areas, genetic drift is more noticeable. Often small (and occasionally large) concentrations of genetic diseases or defects appear in Alpine valleys and in isolated mountain villages. A typical example is albinism—the absence of pigmentation in the skin and even in the retina. Albinos have very white skin and pink eyes and suffer from hypersensitivity to sunlight. The defect is rare (one in ten thousand) but tends to be concentrated in isolated areas here and there. It is the result of single mutations that, instead of being eliminated, have by chance become frequent in specific locations. Elsewhere, albinism hardly exists.

The same applies for other serious hereditary defects, ranging from mental deficiency to blindness or deafness. All of these tend to be concentrated in areas where immigration has been limited. It is sometimes possible to prove that they derive from a single mutant who lived a long time ago. For example, a local form of deafness in the small, isolated Costa Rican village of Taras has been studied in detail. The genealogy of today's sufferers has been traced using church records, and, in fact, researchers have been able to prove that all the cases of the disease descend from one couple who lived some four hundred years ago. Analysis of many deaf patients of Taras has proved that the genetic mutation is always the same.

Field Research in the Parma River Valley, Italy

Genetic drift is responsible for almost all such exceptional situations that occur by chance, and it naturally influences all hereditary features, not just genetic disorders.

In the 1950s, I decided to take a more systematic look at genetic drift in the Italian Parma River valley; this was made possible by a research grant from the Rockefeller Foundation and the determination of my then very young assistant, Franco Conterio (now head of the faculty of sciences at Parma University). We took blood samples in a number of the large prosperous villages in the lowland areas around Parma. We then moved upriver toward the hills (where there are numerous medium-size villages, such as Langhirano, famous for the excellent quality of its hams and Parmesan cheese, and Torrechiara, which boasts one of the most beautiful castles in Italy). Finally, we reached the mountains, where communities number only one hundred or even fewer people. At that time, only a few genes, such as the AB0 and Rh blood groups, could be studied. Our work was made possible by the local parish priests who convinced the faithful to give us blood samples for research. They were almost all seminary students of Don Antonio Moroni, a onetime student of mine and now professor of ecology at Parma and president of the Italian Society of Ecology. We took blood samples in the parish sacristies after Sunday mass.

The theory of genetic drift, which was not yet accepted by all geneticists, told us to expect limited differences among the larger villages down in the valley. The percentage of people in the larger villages with a certain blood group, let us say Rh positive or negative, should have hovered around the 86 percent Rh + , and 14 percent Rh − , typical of this area of Italy. In the hill villages, there should have been greater variations because the communities were smaller. In the mountain villages, we expected to register the greatest variations of all. Our findings fully confirmed our expectations, sometimes right down to the smallest detail.

The Effects of Genetic Drift

Genetic drift is the "random fluctuation of frequencies of the various forms of a gene from one generation to another." It can be reliably predicted using two demographic figures—the number of individuals in each population and the migratory exchanges among them. In the Parma River project, we were able to study the parish records for the whole valley, which show the births, deaths, and marriages in each village from the late 1500s onward. This allowed us to calculate the expected genetic variations, which we then checked against our blood samples.

It may seem strange that the study of the effects of chance can be predicted: surely the workings of chance are unpredictable by definition? In reality, chance prevents us from predicting a single outcome—for example, whether a specific parish will have a greater or lesser percentage of Rh + inhabitants—but it allows us to be as precise as we want in predicting the average variation from village to village if we study enough of them.

The more limited diversification of the lowland villages does not mean that genetic drift ceases to function in large populations. Its effects persist, but it takes longer for them to be felt in a large community.

Eventually, a population with two forms of the same gene—say, Rh + and Rh − —should have only one type or the other. This is true, however, only where immigration cannot reintroduce the eliminated type. If immigration is high, the effect of genetic drift diminishes. Migratory exchanges among peoples are therefore important, too, and these are not necessarily dictated by chance.

Chance and Necessity

The evolutionary factors of mutation, selection, and chance are applicable to every species: it is always a mutation that provides the material (genetic difference) on which evolution acts through chance and natural selection. Natural selection could also be called necessity or destiny. As we have said, mutation proposes but selection decides; chance is an additional factor that literally reshuffles the deck with every deal.

Until recently, chance was thought to be irrelevant. We now know that it is doubly important: its influence is felt both through the statistical effect we call genetic drift, which makes the frequency of different types oscillate from one generation to the next and from area to area, and through the random way in which mutation occurs.

Certain extremely rare mutations can hold great surprises. Every so often, a completely new one, or one that has not occurred in ten centuries or ten million years, appears. This can be totally innovative and positive and can open the way to new evolutionary possibilities. Obviously, a mutation is most likely to spread to the whole population if it is favored by natural selection, but the process of spread is also a matter of chance, or, we can say, luck. Evolution is not just *survival of the fittest*, as Darwin thought. To use the words of Motoo Kimura, who was the greatest expert on the role of chance in evolution, it is also *survival of the luckiest*.

Mutations That Have Made History

The most important genetic change in the history of the humankind has been the growth of the brain and development of new cerebral functions, which occurred from around three million years ago onward.

Extremely important steps have taken place in the last million years. Three hundred thousand years ago, the brain had already reached its present size, which is four times that of our closest cousin in zoological terms, the chimpanzee.

It is extremely likely that more than one genetic mutation was needed to make the brain grow to this extent. The change was also undoubtedly qualitative as well as quantitative. For example, there was major development in the areas of the brain that produce and understand language, which is one of the most significant differences between ourselves and other animals. Other fundamental mutations led to the development of manual skills and the ability to perfect tools that are also exclusive to human beings. This in turn was made possible by mutations that allowed humans to walk on two legs, thus freeing their hands for other activities.

Genetic variation has also led to almost total loss of body hair, which again differentiates us from our closest cousins, the apes. The loss of body hair was less disadvantageous than it would have been for other species—naked animals are, in fact, very rare—perhaps because humans had already learned to wear animal skins by the time the change took place. The harm suffered as a result was limited and the benefits considerable: in the summer heat, it was better not to have fur, and without fur the cooling action of sweat was probably more efficient. During the winter, other animals' furs were available for warmth.

It is true, too, that before furs could be used for clothing they had to be cut and sewn up, skills that form part of the modern human's cultural makeup.

Migration

In addition to mutation, selection, and genetic drift, a fourth evolutionary factor—migration—is generally considered. In reality, the possibilities are endless: migration itself is not a single factor, since it can take on different features and functions.

Virtually every species, including humans, is split into numerous populations living in separate places. If two groups are totally cut off from one another, making all migratory exchange impossible, they tend to diverge genetically. If total isolation continues over a long period of time, they may even become different species; in mammals this takes approximately one million years.

Genetic drift can create profound distinctions between completely isolated human groups. In addition, groups must adapt to their environment, which varies in terms of climate and food sources. Populations living in areas of high humidity have different problems from desert populations, even when they live relatively close by. The diversification caused by genetic drift adds to that resulting from the natural selection that adapts to different environments.

However, two populations are rarely completely cut off: migration diminishes isolation and with it the influence of drift. Migration normally occurs among neighboring or nearby villages and towns. The result is that villages close to one another are always more alike genetically than distant ones. This sort of migration generally continues at the same pace and among the same populations down the generations; an important thrust is the search for a marriage partner, who may not be readily available in a small community. This sort of migration inevitably reduces genetic drift's power to create diversity.

Mass migration also exists. Occasionally an entire group, possibly just a few people at a time, moves and settles in a new area, perhaps a long way away. Migrations of this kind, undertaken to escape famine, natural disaster, war, or simply overpopulation, fill history. Mass migrations have allowed the occupation of entire regions and continents. When a group moved a considerable distance, all contacts between the new settlement and the original territory often ceased, particularly in ancient times when traveling was arduous. Genetic drift and adaptation to new climates then created opportunities for enormous diversification. Mass migration, therefore, leads to greater diversification among groups, while the migration of individuals reduces it.

An Aesthetic Choice

I have used lay language to cover the most important points of a mathematical theory developed by many geneticists in the course of this century. Three of these—two Britons, Sir Ronald Fisher and J. B. S. Haldane, and an American, Sewall Wright—have made the greatest contribution. The centenaries of their births were celebrated recently. I worked with Sir Ronald Fisher for two years at Cambridge and came to know both Haldane and Wright quite well. All three were extraordinary individuals and were rightly considered geniuses. It is a rather exceptional coincidence that they lived at the same time and shared the same ideas and passions. More than anyone else, they have created the mathematical theory of evolution.

I graduated in medicine during World War II. At that time, genetics was virtually unknown in my native Italy. I was attracted to the study of evolution by a consideration that I think of as aesthetic: the sheer beauty of the theory of evolution. My first contacts with the discipline were at Pavia University as a student of Adriano Buzzati-Traverso, the brother of Augusto, Nina, and Dino Buzzati, a well-known Italian writer who is now deceased. In addition to introducing me to the science of heredity and genetics, Adriano also quite unwittingly opened the way to their practical application in my own life by introducing me to his niece, Alba, who was later to become my wife and the mother of my four children.

FIVE

How Different Are We?
The Genetic History
of the Human Species

Can the history of humankind be reconstructed on the basis of today's genetic situation?

This was the question I posed to myself over forty years ago. I made a personal bet that it could be done, because I believed the theory of evolution gives us the key.

Human genetics was an infant science when I started working in research in 1941, and I spent twenty years obtaining the tools I needed to answer it. What follows is the story of how I managed to reach that point, starting from the level of genetic knowledge when I first began.

Between 1948 and 1950, I had a research post at Cambridge University. I was just starting my career and was lucky enough to work with one of this century's greatest geneticists, Sir Ronald A. Fisher, the father of modern statistics and one of the creators of the mathematical theory of evolution. He was a truly exceptional man.

During those two years, I did mostly genetic research on bacteria. I had already studied the genetics of populations and had learned about immunology and blood groups at Milan's serum institute immediately after earning my medical degree.

At Cambridge, Sir Ronald laid great emphasis on blood group research as a means of studying human evolution. He had developed methods of statistical analysis for working on the genetics of ABO and MN groups, and, most important, had interpreted the still very elusive Rh groups, formulating an excellent theory that is most likely still valid today. He had also developed highly innovative research methods that promised to be immensely interesting for the future. Working in this environment, I absorbed both his passion and his thirst for further knowledge and understanding.

Initially, human genetics research dealt almost exclusively with blood groups, then the only fully understood genetic variation between individuals. Basic compatibility rules making blood transfusions possible had been discovered at the start of the century. "Blood groups" are the groups of people who can exchange blood without side effects. There are four main ones: O, A, B, and AB. The O group is also known as the universal donor, because an O person can give blood to anyone, although doctors prefer donor and recipient to have the same group. The ABO system actually comprises three forms (called "alleles) of the same gene, A, B, and O. O people receive the allele (the O form, that is) from both their father and their mother. There are two types of A people, AA and AO. AAs have received A from both their parents (and are homozygotes), whereas AOs have received A from one parent and O from the other (and are heterozygotes—both terms were explained in Chapter 4). Similarly, group B people can be BB or BO. ABs can be only AB.

Many other blood group systems have been found since, but the most important remain ABO and Rh, both of which are essential in transfusions. The rules for blood transfusions are precise. People cannot freely exchange blood because the body reacts to the introduction of heterogeneous, extraneous substances by creating *antibodies*, which attack and eliminate them. The transfusion patient receiving a different type of blood from his own may have serious reactions from antibodies latching onto the blood corpuscles, which agglutinate into bunches. Blood is now unable to flow freely along the blood vessels, causing serious occlusions that can prove fatal.

Blood groups are detected in the laboratory by adding reagents that provoke the agglutination phenomenon.

The studying of ABO groups made it immediately clear that there are different frequencies among different peoples: in Europe, there is an average level of 40 percent O, 40 percent A, 15 percent B, and 5 percent AB. In other populations and within Europe itself, the balance changes.

A new blood group discovered in 1940 has proved extremely important in clinical practice. An excellent New York immunologist, Philip Levine, was able to demonstrate that the death of a particular newborn baby was caused by antibodies produced by the mother to combat a substance present in the child's red blood corpuscles but not in the mother's. The mother's body had responded to this substance with an immunity reaction powerful enough to kill the baby. The substance in question had been transmitted to the child by the father. The substance

MEMBERS OF BLOOD GROUPS

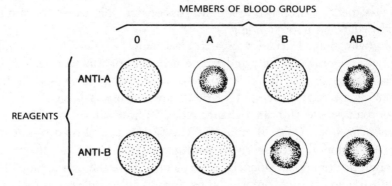

5.1 Agglutination of red corpuscles: ABO blood groups.

has been called Rh, as the result of another discovery made independently in the same year.

The Rh substance was so named because it was first found in the red blood corpuscles of the *Macacus rhesus* monkey but is present also in those of many human beings as well. For example, it is observed in about 85 percent of white Americans (which was the first population analyzed), but absent in the remaining 15 percent. The name Rh is from the first two letters of *rhesus*. People who have the Rh substance in their blood are known as Rh+ (Rh positive); those who don't have it are called Rh− (Rh negative). It was Rh that had caused the death of the child studied by Philip Levine. Both father and child were Rh+ while the mother was Rh−. During pregnancy the mother's body had already begun to react to the presence of the Rh substance in the child's red blood corpuscles by producing specific anti-Rh antibodies. Passed to the child through the placenta, these provoked a fatal case of anemia.

This was later found to be a fairly common phenomenon. Rh− women react to a Rh+ embryo by producing antibodies that agglutinate the baby's red corpuscles and kill the child. The first baby generally survives because the mother does not produce enough antibodies. Subsequent children may be severely harmed or even die before birth. Fortunately, in the Rh system, in contrast to ABO, the antibodies do not exist spontaneously. They are created in Rh− people only in the event of transfusion using Rh+ blood (or in Rh− women by a previous pregnancy in which the baby is Rh+). Nowadays the baby can be saved by a complete blood change immediately after birth.

Further study of the Rh gene has shown that the highest concentration of Rh− is found among peoples of European origin. The average is about 10–15 percent; though in at least one European populations it is higher—the Basques—where it can reach the 30 percent mark. Rh − is rare among Africans, and absent among Asians and the Amerind.

History in our Veins

A Basque hematologist, Michel Angelo Etcheverry, had observed the high frequency of Rh − among the Basques, and had suggested that the Basques of today could be descendants of a proto-European people with a very high incidence of Rh − (perhaps even 100 percent), who inhabited Europe before the arrival of outside populations who were mainly or entirely Rh + . We now have other reasons to believe that this once revolutionary theory is in all likelihood correct.

Another surprising fact—discovered some decades ago–concerns the ABO group and the American Indians, all of whom are O, except for a number of Canadian tribes that have a very high incidence of A (but no B). On other continents, both A and B as well as O are found, with differing concentrations, and the same applies to America's latecomers in prehistoric times, the Eskimos. More baffling still; work on pre-Columbian mummies had suggested that both A and B groups existed among the Amerind several thousand years ago. This research was beset by technical difficulties because certain bacteria produce substances similar to those responsible for A and B. Recently developed DNA testing methods should allow these tests to be repeated to attain a greater degree of certainty. Not all the difficulties have been overcome, however.

There are two interesting explanations for the absence of A and B on the American continent. The first is that the first group of *Homo sapiens sapiens* settlers to cross the Land of Beringia, which joins Siberia and Alaska (and is now submerged and known as the Bering Strait), fifteen thousand years ago or more was very small and contained only group O individuals. This would be a case of genetic drift and a very pronounced example of "founder's effect." We have already described this in relation to islanders like Polynesians: if a founder group is small, it may lack a particular genetic type. That type will then cease to occur unless subsequent mutation or immigration reintroduces it. In reality, the bottleneck need not necessarily be among the founders. It may come later, which is how we can perhaps explain the presence of A and absence of B in some northern tribes.

The second possible explanation is that natural selection has made the other groups disappear, for example, because the O group turned out to be more resistant to certain diseases. In this case, too, an additional explanation would be needed to explain the presence of A in the far north.

The lack of groups A and B throughout most of America has been linked to a disease that appeared in Europe a very short time after Christopher Columbus's return: syphilis. An attempt has been made to see if there is a link between the O blood group and resistance to this disease. Syphilis was widespread until World War II, but it has now been virtually wiped out thanks to the discovery of penicillin, which is much more effective than the previous treatments used. Even today, in the United States, couples must take a blood test, the Wasserman, before marrying to ensure that they are clear of the illness; if affected, most postpone the wedding until after successful treatment. This law was designed to prevent sufferers from infecting their partner and even their children.

Group O individuals were found to be less prone to syphilis, but they responded more quickly to the treatment available before penicillin. This suggests that O types had stronger resistance and that the lack of A and B in the Amerind is due to natural selection, since their resistance was less marked. It is a possibility, not a certainty.

People belonging to certain ABO groups are particularly prone to various problems of the digestive system (ulcers and duodenal tumors) and certain infectious diseases, including tuberculosis, streptococcus conditions, and others caused by strains of *Bacterium coli*, which are particularly common in children. Therefore, natural selection probably has some influence on ABO types. Despite this, with the exception of the Amerind, the frequency of ABO groups around the world is quite stable. Variations are much more limited than for other gene types. Perhaps natural selection, while favoring different individuals in different climates and regions, tends to maintain a fairly constant balance for certain genes and circumstances. This happens when people receiving different genes from their parents (heterozygotes) have a selective advantage over others (as in the example of thalassemia, in the presence of malaria). For some reason, this selection mechanism did not work with the Amerind, and the A and, particularly, B types have disappeared almost everywhere in pre-Columbian America.

Modern work on mitochondria in Amerind people seems to indicate that the number of founders was not very high, but this gives us no idea of how many there actually were. The two theories, natural selection and genetic drift, are not mutually exclusive, and for the moment the question remains open.

How Can We Trace the Past?

Blood typing was about the only scientifically rigorous tool for studying genetics when I began to look at human evolution. It was not enough to trace the history of humankind. With only the ABO and Rh groups to go on, the only interesting statements possible would necessarily have been very generic and limited to isolated cases, such as the Basques.

My supposition was this: if enough data on a number of different genes are gathered, we may eventually be able to reconstruct the history of the entire human species, i.e. its evolutionary, or phylogenetic tree. The accumulation of large quantities of data probably would give us a clearer understanding of the vague conclusions reached, sometimes intuitively, using a single gene. I developed a method of analysis that would allow me to use data on the genetic differences between peoples, were sufficient information on many genes in different populations to become available. It would have been too difficult for me alone to analyze the indigenous populations of various parts of the world, but I was convinced that the information would accumulate fairly rapidly, thanks to the large number of researchers working on blood groups and other genetic variations being discovered all the time. I felt that it was just a question of time.

About 1960, scientific literature seemed to provide enough material to begin this work. I had both American and Italian research grants at Pavia University, and I invited Anthony Edwards, another of Fisher's students to join me. Anthony, who is now back at Cambridge, is a great expert on population genetics and statistics, and also on data processing. Thanks to a period in which the normally penny-pinching Italian Ministry of Education wanted to encourage data processing in university institutions, we had a new Olivetti computer at our disposal. This was a new tool for everyone, and for a long while we had it almost entirely to ourselves. The computer filled a large, air-conditioned room, but was much less powerful than a modern PC. Anthony wrote a number of programs for the application of the most suitable statistical methods to our body of data. Nowadays, these sorts of programs are widely available in easy-to-use packages, but at the time you had to write your own. Others we made were applicable to our specific problem of reconstructing the phylogenetic tree, and we developed a number of different methods of reconstruction other than the first.

Many genes exist in several different forms, but their proportions may differ very little from one population to the next. Fortunately, though, there is a significant number of genes, that do differ to a greater degree among populations, and these are generally more helpful. To trace the history of humanity, I needed to be able to base our analysis on a great many differing genes.

We needed to find a way to obtain a single number to express in some suitable, synthetic way the global difference between two populations on the basis of all the genetic information available. We call this figure "genetic distance." Taking Rh + / − as an example, if we register about 20 percent Rh − negative among all Basques, 15 percent among the English and 2 percent among the Chinese, then a simple genetic distance could be 5 percent between Basques and English (20 − 15 = 5), 18 percent (20 − 2 = 18) between Basques and Chinese, and 13 percent (15 − 2 = 13) between English and Chinese. In reality, we chose a more complex formula. We subsequently learned that the exact formula used to calculate the difference is not particularly important. However, it *is* necessary to use as many genes as possible: whatever the underlying formula used to calculate distance starting from a single gene, the overall genetic distance will be the average of those found for every gene.

An Evolutionary Tree Based on Blood Groups

During 1961 and 1962, we brought together data published on about fifteen populations, three per continent, dealing with a total of twenty genetic variations. All the data were on blood groups: ABO, Rh, and three other systems (known, for the books, as MN, Diego, and Duffy).

We assessed the genetic distance between pairs of populations on the basis of these data, for each of the 105 combinations offered by two-by-two comparisons. This gave us the most rational tree possible using the available data and applying our specially developed reconstruction methods.

Our first attempt is still approximately correct today, despite the relatively small number of genes used. Populations from the same continent tended to fall close together, and this was a good sign, because it was reasonable to expect that populations from the same continent should form closely knit clusters. Some populations gravitated around the same branch: the American Indians, for example, turned out to be related to the Eskimos and more distantly to the Koreans. This was another encouraging sign, because there is wide agreement that the

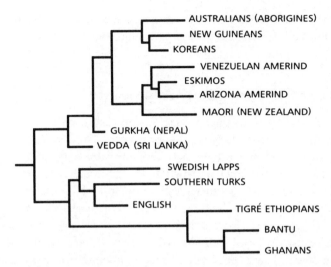

5.2 Evolutionary tree drawn up in 1962 by Luca Cavalli-Sforza and Anthony Edwards on the basis of five blood group systems from fifteen populations, three per continent.

Amerind and Eskimos are both Mongolian in origin, and that they arrived in America from eastern Asia via Siberia and Alaska.

The scanty knowledge available at the time meant that some of the results were hard to assess. For example, the Europeans were a sort of intermediate group between Africans and Asians. The extreme poles of human variation were represented by the Africans at one end and the New Guineans and Australian Aborigines at the other.

Anthony and I were naturally rather satisfied with this first effort. By plotting our tree on a geographic world map, we seemed to obtain an idea, albeit approximate, of the routes taken by modern humans in their expansion. We now had some reason to believe that it really was possible to trace the history of living populations using a mathematical method. We were winning our bet, but we knew that this was only a first step. We still had a long way to go.

The forks of the tree in Figure 5.2 can be expected to correspond in historical terms to the separation between two peoples; to the moment, that is, when a group broke off and moved far enough away to severely reduce or interrupt all contact with the parent population. If the reconstruction worked, the sequences of branches should have corresponded to that of the splits, and if we were very lucky, the position and length of the branches would correspond to the time in which the splits occurred.

We knew we had to wait until a later stage, when we would have much more solid data, before accepting our conclusions in detail.

A Tree of External Body Appearance

Our next job was to check these results using completely different data, for which similar information also existed. These were hair and skin color, stature, and the other characters measured by anthropologists, usually called anthropometrics, such as the circumference of the thorax, length of limbs, and the anthropologists' much-loved skull measurements (including the transversal and front-back diameters). The calculation of the ratio of these last two quantities led to the so-called cephalic index, and the distinction between dolicocephalics (long-heads) and brachicephalics (wide-heads). This type of measurement became popular in the middle of the last century, but today is considered of little biological or evolutionary interest. These features of the external aspects of the body we know today, are dictated only in part by heredity; the environment in which individuals grow up also plays a very important role in determining our body size and shape.

Using the same method of tree reconstruction we had applied to genetic distances, anthropometrics gave us a rather different tree. The Australian Aborigines now seemed closer to the Africans than to the Asians (Chinese and Japanese), while the Amerind, who were related to Chinese and Japanese in the genetic tree, were now more similar to the Europeans.

There were two possible explanations, a short term, and a long term, environmental effect. The first had already been shown by an American anthropologist, Franz Boas, at the beginning of the century. People who migrated, still young, to the United States (usually from poorer countries), showed a substantial increase in their physical measurements when compared with their relatives who stayed home. We notice a similar phenomenon today with the height of old and young Japanese: the latter are definitely taller, on average. Both phenomenon are most likely due to a recent change in environment, which has a strong short-term effect. Probably diet, in particular vitamins, and other not well-understood factors, are responsible for it. But environment certainly has another effect, because different climates can affect our physical constitution in the long term through natural selection. It is well known that, in animals, body size and proportions show correlations with climatic differences, representing genetic adaptations to climate conditions. But the distinction between the two environmental effects, short- and long-term, is not as easily seen as one might hope.

Which Features Reveal the History of Humankind?

It was obviously necessary to explain the discrepancy between the genetic tree and the anthropometric one. We weren't worried, because anthropometric features are not as strictly hereditary as blood groups. We knew that our results could be strongly influenced by both short- and long-term environmental factors. Height and all the other closely related body measurements depend in part on lifestyle and other environmental conditions: those who eat more and better, grow bigger. Skin color is influenced by exposure to the sun (even Africans can get a tan.) Naturally, genes also influence body size, skin color, and face shape. If, however, the genetic component of these features is high, why was our anthropometric tree different? Were we to conclude that the pressure of genetic influence on anthropometric features is small or nonexistent, and the differences are of a short-term nature? In fact, the importance of genetic factors in determining these features is not fully understood. We need not overconcern ourselves, though, because it is very likely that there is a strong long-term effect.

External body features, such as skin color, and body size and shape, are highly subject to the influence of natural selection due to climate. It is reasonable, therefore, for the Australians and Africans to be close on the anthropometric tree because they live in similar climates. In reality, it is risky to use these features to study genetic history, because they reveal much about the geography of climates in which populations lived in the last millennia, and little about the history of fissions of a population. The anthropometric tree tells us that both the Australians and New Guineans have long lived in warm climates and the Mongols in cold ones, but it will not help us discover when these peoples separated, nor from which preexisting peoples they descended. Subsequent research revealed that climate can also have a bearing on blood groups and some of the other genes in our research sample. This is probably also the work of natural selection, although the results are much less emphatic than for anthropometric features.

Casual events have considerable bearing on gene variations. If interpreted correctly, these variations can reveal the background to the splits accompanying the settlement of new regions and continents. It may seem odd that chance events can help reconstruct the history of evolution in terms of the sequence of fissions and migrations of populations. The reason—as we saw in Chapter 4—is that chance phenomena are perfectly predictable if we calculate averages from a large enough body of data.

It is interesting to note that Darwin, who knew nothing of genetic drift or random mutations simply because they had never been thought of, recognized that the features most useful when studying evolution are the ones he calls "trivial." Today we call them "selectively neutral" because they are not adaptive. We now know that many features are entirely or virtually neutral in terms of selection, exactly as Motoo Kimura first argued in a famous 1968 article on the interpretation of molecular evolution data. A long debate ensued between Kimura with his neutralist position and other biologists who were initially skeptical about Kimura's revolutionary proposals. The question has now been settled to the partial satisfaction of both parties. Modern biologists have grown up with the idea that natural selection is the only evolutionary force that allows adaptation to the environment and, therefore, survival. This is true: both DNA and proteins show unequivocal signs of natural selection. There are numerous other signs, however, of the importance of chance, particularly in those areas of DNA safeguarded from the pressure of natural selection because they are silent. For example, the genome holds many duplicates of active genes, which are unable to work as a result of mutations that have rendered them inactive. We call these *pseudogenes*. Natural selection is powerless against this DNA, which is especially precious for our work for the very reason that over long periods of time the changes affecting them have been random. Our survey data did not include pseudogenes, which can be studied easily only today and are relatively rare, but various analyses have led to the conviction that most of the genes we studied are influenced to a broad extent by genetic drift and therefore by chance, perhaps even more so than by selection.

In deciding which of the two trees is the best representation of human's evolutionary history, we concluded that it was the genetic tree, which can tell us more about the history of descent, i.e. of common ancestry, while the anthropometric data told us about climate, which was interesting but not the type of history we were looking for.

Genes and Anthropometric Features

Subsequent developments proved us right. Among these is the work done by the well-known Harvard anthropologist W. W. Howells.

Howells traveled the world measuring numerous skulls in seventeen populations in different continents. His methods of analysis were similar to ours for blood groups. Anthropometric measurements often vary according to the measurer, and a considerably higher degree of precision is attained if the measurer is always the same person. The evolutionary tree Howells obtained from his crania is practically the same as ours for body features (which took account of skull measurements, as well as other somatic features and skin color). Howells's tree puts Africans and Australians together, meaning that if the diagram has significance in evolutionary terms, the two groups must have a recent common origin. The same happened with Amerinds and Europeans, a result most similar to that of our anthropometric tree.

We have said that we rejected the tree based on external features because they are mainly the result of adaptation to environment and tell us more about climate geography than about evolution by descent. The same objection applies to Howells's head measurements: we demonstrated that his most important data were all closely related to climate.

This is not really surprising, because all body dimensions adapt to climate; we have already discussed this at length in relation to the pygmies and to modern humankind's settlement of the Artic. Wider observation, for example, points to a precise relationship between stature and climate (expressed in terms of average annual temperature). In cold

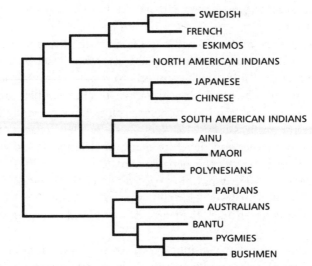

5.3 W. W. Howells's tree of world populations based on skull measurements.

climates, it is better to be large in order to limit the body heat lost through surface area; in hot climates, the exact opposite is true (see Figure 1.4). On average, a tall person has a larger cranium than a short one, and all other body dimensions tend to follow the same rule. All head and body measurements tend to be correlated between themselves and to temperatures.

In subsequent work, Howells calculated statistical indices that reflect head shape rather than size. This too turned out to relate to climate. The Mongol head, with its wide, flat face, is a sign of adaptation to the cold; because a face without jutting parts is defended more easily. In warm climates, pointed features and a long face are useful. This is the typical African shape and not so very different, either from that found in the tropical areas of Southeast Asia and New Guinea.

Face shape tells us the same things as head measurements. External body features mainly express the results of natural selection due to different climates. There are also statistical reasons why body measurements are less helpful than genes in the analysis of evolution.

Polishing the Evolutionary Tree

I have shown our first tree (Figure 5.2) because, despite the small amount of data available and even though it needs some retouching, it was quite satisfactory. Today we would have fifteen times as much material to use.

Over the years we continued to elaborate on this tree, adding new information and correcting the inevitable mistakes. Data on hundreds of genes are now available: much continues to relate to blood groups; much comes from new sources, such as the more recently studied proteins and enzymes. The genes useful to studying evolution are generally known by the collective term genetic markers, because they change from one individual to another and therefore mark, or distinguish individuals and populations in terms of their hereditary material.

In the last decade, it has also become possible to work directly on DNA, at a molecular level. This allows us to make much more complete and direct genetic analyses than use of only blood groups and proteins did.

The results obtained continue to be very similar, although areas of doubt persist. We can resolve these only by accumulating more data and testing the various possible explanations. A more recent tree shows the

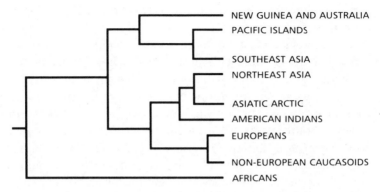

5.4 Genetic tree of the world's populations based on 110 genes, all of which are blood groups, proteins, enzymes, and so on. Drawn up by Paolo Menozzi, Alberto Piazza, and L. Luca Cavalli-Sforza (From *History and Geography of Human Genes*, Princeton Univ. Press, 1984).

results of a survey of 110 genes in forty-two native populations around the world. We employed classic gene types of the kind used until a few years ago for this tree. Our sample included blood groups, blood proteins, enzymes, and other hereditary features. The full tree is given in Figure 7.7; in Figure 5.4 above we give a simplified version grouping the forty-two populations into nine major categories.

The biggest difference in the tree is between Africans and non-Africans, once again reinforcing many paleoanthropologists' view that modern humans originated in Africa and later spread around the world. Our initial analysis didn't reveal that the first fork separated Africans from non-Africans. This particular discovery was made by geneticist Masatoshi Nei. The limited blood group data then available indicated a marked similarity between Africans and Europeans, and, as a result, the first fork tended to correspond more closely to a split between Afro-Europeans and the rest of the world. This was in part a "statistical" error, as we didn't have enough data to be absolutely accurate and chance intervened to make our tree differ in detail from present results.

Non-Africans sit on two major branches. One carries today's inhabitants of Southeast Asia and the populations who most likely reached Australia, New Guinea, and the Pacific Islands from there. The group on the other branch populated northern Asia: most headed eastward (into Siberia and then America), and the rest (mainly Europeans and non-European Caucasoids) headed westward. The Caucasoids are mainly fair-skinned peoples, but this group also includes the southern Indians, who live in tropical areas and show signs of a marked darkening in skin pigmentation, although their facial and body traits are Caucasoid rather than African or Australian.

In this tree, the inhabitants of Southeast Asia tend to fall with those of Australia and New Guinea. This positioning is not absolutely certain, because slightly different approaches indicate that the Southeast Asiatics ought to be grouped with the Mongoloids who live farther north rather than with the inhabitants of Oceania. There are genetic variations among the peoples of Southeast Asia that the information gathered to date does not adequately explain. Certain groups, such as the Vietnamese and some Cambodians, are more Mongoloid in type and nearer to the Chinese or Japanese; others, such as the Malaysians, and the "negritos" in particular, look more like the peoples of Oceania. Further research will undoubtedly help resolve these doubts. There is a general need to increase genetic data on human populations, and the new molecular genetic techniques are proving extremely useful. This is the purpose of a recent international program to study the diversity of the human genome, which I and many colleagues have been working on since 1991.

Reciprocal Migrations

The main complicating factor in this scenario comes from the frequent and reciprocal migratory exchanges between neighboring peoples that take place after the splits corresponding to the forks in the genetic tree. Many of the problems thus created have so far been resolved only in part.

Large-scale exchanges give rise to particular anomalies. We observed one among the Europeans, and more broadly, among the Caucasoids in general. The complicating factor is that the Caucasoids (a group including Europeans, inhabitants of the Middle East, Iranians, Pakistanis, and Asiatic Indians) resemble both Africans and Asians. One consequence that emerged even in the first tree is that the Europeans, and in part the Asians too, have shorter branches. One possible explanation is the existence of genetic exhanges between Caucasoids and Africans, on the one hand, and Caucasoids and Asians, on the other. Since the Caucasoids are sandwiched geographically between Africa and eastern Asia, it is reasonable to believe that migratory exchanges took place in both recent and not-so-recent times.

There is, however, another possible explanation, of a technical nature: the selection of markers used in research to date has mostly involved variables identified among populations of European origin. This has been a question of convenience, not racism or wanton neglect of certain groups. The fact is that researchers needing blood samples from individuals and families to control the hereditary nature of recently distinguished markers naturally turn to the donors closest to hand, who (also in the United States) are, or were until recently, almost always of European origin. New markers that take account of the variations in non-European populations need to be found. This is now happening at an increasing rate.

When Did the Races of Humanity Separate?

The most definite dates, in archaeological terms, relate to the occupation of new continents. We now possess four fairly reliable dates, although future findings could change them.

The first date refers to the oldest modern humans found in both Africa and the Middle East around one hundred thousand years ago. Available dates do not distinguish clearly which is older, but earlier skulls from Africa appear to show signs of a trend toward modern human forms, strengthening many archaeologists' conviction that *Homo sapiens sapiens'* birthplace was in Africa. The presence of modern human sites to the west and east of Suez one hundred thousand years ago suggests that the journey from Africa into Asia (or, less probably, vice versa) occurred at about that time, so that the first diversification between Africans and non-Africans would have taken place then or a little beforehand. The date of "African Eve" is earlier, but this is not a contradiction, as it refers to a different event: the generation of the earliest recognizable mutant, which marks the last common ancestress. As we mentioned earlier, the separation of future non-Africans and their immigration out of Africa, which is the fission of the tree, must be later.

The first human vestiges in Australia and New Guinea have been dated at fifty-five to sixty thousand years ago. The genetic distance between the Oceanian Aborigines (a term for the slightly diverse Australian and Papuan Aborigines) and their Southeast Asian neighbors is about half that between Africans and non-Africans. The date of entry into Oceania (50–60,000 years ago) is about half that of the finding of modern humans in both Africa and the Middle East (100,000 years ago).

The other two dates are more recent and indicate the times of the occupation of Europe (probably from western Asia, around thirty-five to forty thousand years ago), and America. The date for the latter is still rather unclear, unfortunately, but it almost certainly falls between fifteen and thirty-five thousand years ago.

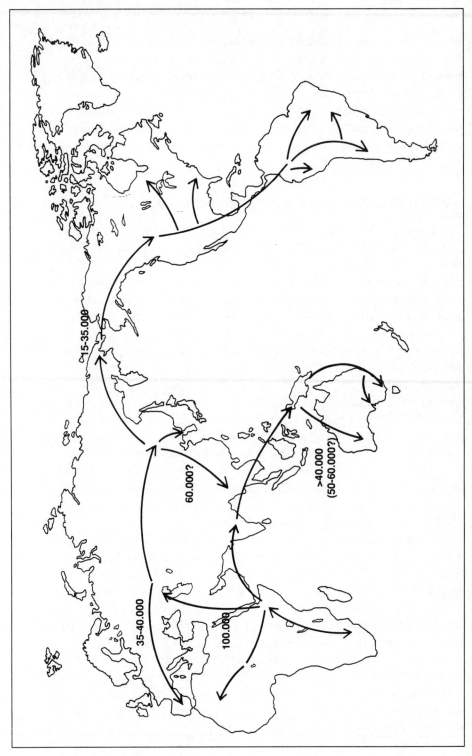

5.5 Map showing the probable expansion routes of anatomically modern humans (*Homo sapiens sapiens*) from their original birthplace in Africa to other continents, including the probable dates of arrival.

Figure 5.5 shows the probable routes taken by modern humans (although the scarcity of finds leaves archaeologists with only a hazy idea of the exact itinerary) and the dates of their arrival in various continents, established using archaeological data.

At this point we can calculate the relationship between the date of arrival on a certain continent and the genetic distance between the two resultant groups who share the same ancestors: the emigrants, and those who stayed behind. The genetic distances corresponding to the forks of the tree of nine populations in Figure 5.4 are given below. The index of the four comparisons of genetic distance among human populations sets the maximum distance known, that between Africans and non-Africans, as 100:

Separation of Peoples	Date	Genetic Distance
Africa and rest of world	100,000 years ago	100
Southeast Asia and Australia	55–60,000 years ago	62
Asia and Europe	35–40,000 years ago	48
Northeast Asia and America	15–35,000 years ago	30

Genetic distance among peoples must increase with separation time; at its simplest, it will increase at a constant rate. The table shows a precise progression: as we expected, the smaller the period of time since separation, the smaller the genetic distance. Unfortunately, the dates are approximate and, although we took averages from 110 genes, the level of statistical error remains high (around 20 percent). Taking account of statistical error, the first three comparisons agree well, as if genetic distance really does increase regularly and proportionately to separation dates. The last comparison is too imprecise to be reliable, although we can use it to try to calculate America's date of occupation, starting from genetic distance and using the first three comparisons as our basis. The date obtained in this way is thirty thousand years, which falls well within those suggested by archaeologists (fifteen thousand years ago at the latest, thirty-five thousand at the earliest) and is clearly closer to the earliest limit.

Differences That are Skin-Deep

The differences between us are very small. Accustomed as we are to noticing variations in skin color or facial structure, we tend to assume that the differences between Europeans, Africans, Asians, and so on must be large. In reality, the genes responsible for these differences are those that have reacted to climate. All the peoples now living in the Tropics and the Artic *must* have adapted to local conditions in the course of evolution; large individual variations are not permissible in features that determine survival in a set of surroundings. We must also bear in mind that the genes that react to climate are those that influence *external features*. Adaptation to climate for the most part requires changes of the body surface, because this is our interface with the outside world. *It is because they are external that these racial differences strike us so forcibly, and we automatically assume that differences of similar magnitude exist below the surface, in the rest of our genetic makeup. This is simply not so: the remainder of our genetic makeup hardly differs at all.*

We cannot study in depth the genetic difference between races in terms of skin color and other aspects of appearance because we have not yet identified exactly which genes are responsible for them within our DNA. We only know that at least three or four different genes combine their effects to determine the extremes of skin pigmentation, such as black and white. For any other gene affecting our external aspect, we know even less.

How Different Are We?

If we ignore skin pigmentation, the differences between races can be considered purely quantitative and not qualitative, in the sense we never find two races, A and B, that are totally different, not even for one gene. We can draw up a frequency table for three of these genes (see p. 125). The first, GC, has two major forms, GC-1 and GC-2. GC is a blood protein that binds vitamin D and regulates its distribution within the body. The table below shows percentages of GC-1 and GC-2 types in various regions. Because only two forms of this gene interest us for the moment, the sum of their percentages is always 100. The first two rows of the table illustrate how little the proportions of the two forms vary among populations.

The other two genes shown are HP and FY. HP is another blood protein that binds the hemoglobin liberated by the red blood cells when they decay spontaneously at the end of their natural life or when they are destroyed by a disease, such as malaria. There are two major forms, HP-1 and HP-2. Very occasionally, the protein may be missing. This is known as HP-0; it is compatible with survival but has drawbacks for the carrier. We give the percentage of HP-1 only, since HP-2 is almost always 100 minus HP-1.

The last gene, FY-0, is the absence of the FY substance. This is normally found on the surface of red blood cells and facilitates their acceptance of a particular malarial parasite, *Plasmodium vivax*, which, like all malarial parasites, multiplies in the red blood cells. The absence of the protein makes it hard for the parasite to multiply, and therefore grants the carrier a certain amount of protection against this strain of malaria.

Gene Type	Europe	Sub-Saharan Africa	India	Far East	South America	Australia
GC-1	72%	88%	75%	76%	73%	83%
GC-2	28%	12%	25%	24%	27%	17%
HP-1	38%	57%	17%	23%	60%	27%
FY-0	0.3%	87%	3%	0%	0.2%	0%

The fluctuation of GC-1 or GC-2 among peoples is minimum, from 72 to 88 percent, (16 percent at most). As said, GC is probably important for the normal circulation of vitamin D in the body. It has been suggested that the GC-2 form is more suitable in areas where sunlight is intense, and GC-1 in areas where it is less so, but the difference must be limited because the variation range is very small, even between climatic extremes.

The HP-1 gene varies more, its maximum range being 43 percent (from 17 percent to 60 percent). The HP-2 percentages calculated from the levels of HP-1 naturally show a corresponding variation (from 83 percent to 40 percent). Clearly the GC gene varies somewhat less than HP.

FY varies most: from 0 percent to 87 percent. The FY-0 type has a selective advantage in Africa, where a malaria parasite is common. As a result FY-0 is very common in Africa. Elsewhere the type hardly exists at all. The remaining genes out of the 110 we examined vary much less on average. FY displays a difference between Africans and non-Africans comparable to those we believe exist for skin color between tropical peoples and those who live in areas very distant from the equator.

The differences between "races," a term whose meaning and limitations we will examine later, are therefore very limited and quantitative rather than qualitative. Within continents, the differences are, on average, even smaller. Seen in this light, the confusion, misery, and tragic cruelty caused by racial differences between humans are, to use the words of Shakespeare's Macbeth, "a tale told by an idiot, full of sound and fury, signifying nothing."

SIX

The Last Ten Thousand Years
The Great Trek of the Cultivators

Ten thousand years is the period of time separating us from the start of a veritable revolution in the history of the human race: the shift from a hunting-gathering economy to the direct production of food. Humans had previously lived on what they found in the wild. Over millions of years, their hunting ability and understanding of the environment had developed to an extraordinary degree, allowing them to amply exploit the opportunities offered by their surroundings.

The evidence left by our antecedents of fifteen to twenty thousand years ago in Europe suggests that they had a high standard of living. These people hunted, fished, and gathered enough plants, fruit, and roots to support small communities and survive well; even today their art, ornamental objects, and tooling skills inspire our admiration.

Some ten thousand years ago, however, these people started to produce their own food, by cultivating plants and rearing animals, generally the ones already eaten in the wild. This led to an enormous increase in the potential numbers of people the earth could support. In the four to five hundred generations since then, the world's population has increased over a thousandfold, from a mere few million to today's six billion, a figure set to grow further still.

On the Trail of the Megaliths

Why would a geneticist be interested in food production? My interest was sparked in a somewhat roundabout way.

About thirty years ago, I happened to visit the Pigorini Prehistoric and Ethnographic Museum in Rome. Sometime earlier I had been to Sardinia and seen the Nuraghi, the large dry-stone buildings scattered in large numbers across most of the island. The Nuraghi are extraordinary, and are hard to imagine without actually seeing them. They are towers that probably served as both dwellings and fortresses, laid out in an intercommunicating network that covered the whole island. Building of this kind commenced about 3800 years ago and continued for at least one thousand years. About six thousand Nuraghi remain today, demonstrating that Sardinia must have been densely populated in that period, with perhaps two to three hundred thousand inhabitants. Today, there are only about ten times more.

During this museum visit I discovered that in Puglia, in southern Italy, there stand a large number of monuments very similar to the Nuraghi. These "specchie," as they are known, have been largely destroyed, unfortunately. I subsequently discovered the existence of not dissimilar monuments on other Mediterranean islands; there must have been a single civilization that reached these places and built the monuments.

In fact, large prehistoric stone buildings are found along a strip of territory reaching from the Atlantic Ocean to India and almost as far as Japan. The most important instances, and highest concentration are found in various parts of Europe, however, generally near the coast. In reality, the Nuraghi are just one example. These large buildings had various architectural forms and uses—dwellings, burial chambers, or temples. They may all have been erected by a single population of colonizers, navigators, and cultivators, which we know as "Megalithic," for want of a better term. This population may well have originated in France, Britain, or Spain, since that is where the oldest monuments have been found. The most famous is Stonehenge, west of London, which has been rather imaginatively nicknamed a "stone computer" because it was probably an astronomic observatory of some sort, used to predict important crop-sowing and agricultural phenomena.

The Sardinian Nuraghi are architecturally different from the buildings erected elsewhere. The specchie in Puglia appear to have been built in a defensive network, like the Nuraghi, and their shape, (when not reduced to mounds of stones) also is similar.

My reaction on noticing these similarities in building types was to think that there ought to be genetic similarities, too, among the peoples building this particular kind of megalithic monument. The job should have been facilitated by the fact that the Sardinians were already known to be genetically very different from almost all other Europeans, probably as a result of genetic drift during long periods of isolation. Sardinia, a large island some distance off the Italian mainland, has always seemed cut off, and its people have developed, to some extent, independently of the rest of the Mediterranean. But little was known of the genetics of the

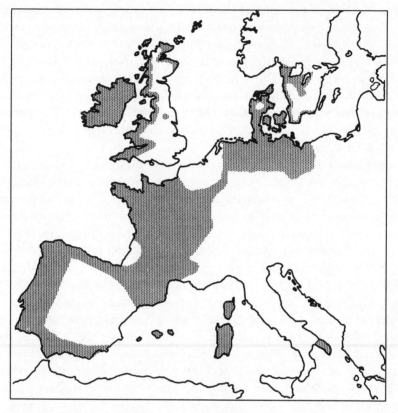

6.1   Distribution of megalithic monuments in Europe.

people of Puglia, so I decided to organize a small expedition with two or three like-minded colleagues to collect blood samples from the most interesting areas. We planned to make genetic comparisons initially with the Sardinians, and perhaps later with other populations.

A False Start and Some Intuitive Thinking

Our first attempt to establish a link between archaeology and genetics was an out-and-out failure. The results showed clearly that there is no particular resemblance between the Sardinians and the people of Puglia, who are, in fact, exactly like the other inhabitants of southern Italy. This failure taught me an important lesson: cultural similarities alone are unreliable indicators of genetic similarities.

The existence in Puglia and elsewhere of monuments similar to the Sardinian Nuraghi ought to mean that a people able to build this kind of construction came and either established good relations with the local inhabitants or imposed their rule through armed force. In any case, this people convinced or coerced the native population into helping them erect their huge buildings, which were used for habitation and defense and perhaps for religious, political, and even astronomical purposes as well. Rather than colonizers, the Megalithics may have been a priesthood or some kind of prehistorical aristocracy, who had good ships and perhaps good weaponry, as well as a much more advanced understanding of astronomy and architecture than their contemporaries. They imposed much of their culture on those they encountered, although they were probably not very numerous compared with the relatively high density of farming peoples who had already colonized the Mediterranean coast.

So the contribution of megalithic genes remained limited and didn't alter the genetic makeup of the peoples encountered. In cultural terms, however, they left an imposing legacy, and these grandiose monuments remain one of the great mysteries of prehistory.

Other interpretations could be made—for example, that other, more numerous and important migrations succeeded the arrival of the Megalithics, which might have diluted the latter's genetic contribution to the extent that it is undetectable with our methods. So far, however, there has been no clear indication in any region of Europe where these monuments are found of a genetic inheritance attributable to the Megalithics.

The Importance of Numbers

If we knew the history of humankind or if we could look into a crystal ball and see all that previous generations have done and been, we could see that genetic and archaeological data are part of the same story. Since we know little of our past, and the sciences that study it often provide separate (and noncommunicating) fragments of knowledge, it is important for them to learn how to help each other.

I started to analyze the development of agriculture (which was the cause of the first demographic boom) with the belief that it would lead to a closer relationship between archaeological and genetic phenomena. In fact, when the number of individuals that form a population increases drastically, forcing a fairly large section to occupy new areas and introduce their genes there, the resulting genetic makeup of the mixed population depends exclusively on the numerical ratio between newcomers and natives. Very straightforward calculations can be made on the basis of these two variables: immigrants and residents.

Agriculture's Dawn

Evidently, no one can tell how big the contingencies of emigrant farmers were. They were, however, larger than in other migratory phenomena. So I started from the archaeological period in which agriculture began, which in Europe is called Neolithic, or New Stone Age, because of the innovations in the cultivators' stone tools. These are now designed for several new purposes and are often skillfully filed into shape, rather than just chipped like the Paleolithic or Old Stone Age instruments used by their hunting predecessors.

The cultivators needed new kinds of working instruments such as scythes to harvest. They worked obsidian, where it was to be found, because obsidian provides a very fine cutting edge. Neolithic farmers used it to make sets of identical blades for the edge of the scythe. These cut well, and the hardness of the stone meant that the blade remained sharp for a long time.

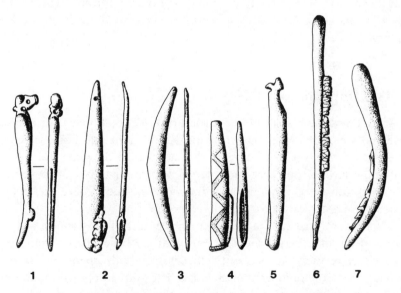

6.2 Neolithic agricultural tools: bone instruments with flint blade: (1–5) reaping knives from Iran and Egypt; (6–7) small sickles from Bulgaria and Spain.

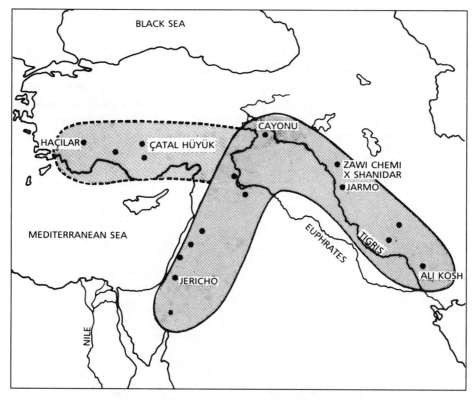

6.3 The fertile crescent in the Middle East, and its extension into Turkey, with an indication of the most important traces of the domestication of cereal crops and animals.

I now joined forces with a young archaeologist, Albert Ammermann, who worked with me first at Pavia University in Italy and then later for several years at Stanford. Together we examined what was known about the spread of agriculture from its various birthplaces, the best known of which was then the Middle East.

The domestication of cereals began in the Middle East. Wild cereals were already a source of food, but agriculture allowed greater control of both the quantity and the quality of the yield, as well as the location of supplies. Various types of wild wheat and barley were also cultivated and livestock, particularly sheep, goats, cattle and pigs, was kept.

A very powerful mixed economy grew up this way. The greater availability of food allowed people to have many more children, and to build large villages and the first small cities. A fine example of an early city is found in Anatolia in Turkey, at Çatal Hüyük. The city rests on an artificial hummock created over the millennia by a string of human settlements. The first layer of clay dwellings dates back about ten thousand years. When these became uninhabitable, new ones were built on top, and so on until the hummock was formed. The site has been excavated with care, and this has allowed its extreme antiquity to be established. More than nine thousand years ago, it already housed a farming community of about five thousand people. Wall decoration had developed, and probably fabric decoration, too. Patterns similar to ones discovered at Çatal Hüyük can still be found on modern ornamental fabrics and carpets, such as the Anatolian kilim.

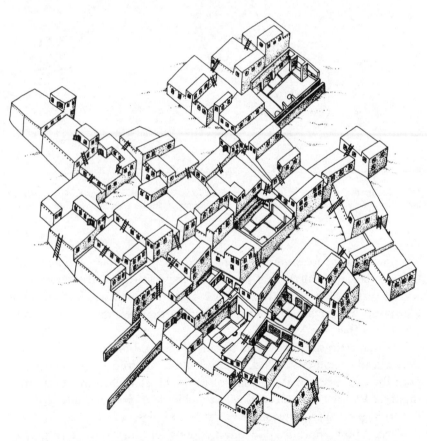

6.4a The first Neolithic city at Çatal Hüyük, in Turkey, a hummock created by numerous layers of debris, each of which is the archaeological remains of a city built on the ruins of the previous one. There are no roads (access was across the rooftops).

6.4b Çatal Hüyük. Examples of decoration found in rooms probably used for religious purposes. The patterns have come down to us and are still recognizable, despite considerable stylization, in the carpets and kilim of this and neighboring areas. Recurring themes are bulls' and rams' horns, fertility goddesses, vultures, and caves.

Demographic Explosion

The first effect of agriculture within a region was therefore the possibility of providing nourishment for far more people, which in turn led to growth in population density. Habits and customs determining birthrate are always deep-rooted. Before agriculture developed, these customs kept population growth at a low level. Agriculture made higher birthrates both feasible and desirable. Once it begins to rise, birthrate does not easily slow down again.

The hunter-gatherers of earlier times were presumably like those of today, who have an average of five children, one about every four years.

A four-year gap means that the parents can always carry the youngest child on their backs or in their arms, while the older ones are already able to walk at a reasonable pace. Longer gaps also mean that children can be breast-fed until the age of three, and this in turn lowers the probability of another pregnancy. An average of five children per woman keeps the population substantially stable, because more than half will generally die at an early age and in any case before reaching adulthood. Of a couple's children, therefore, only two tend to procreate, so population remains stable or at most increases slowly.

Farmers, however, have no reason to limit the number of children. They are settled and therefore don't have the problem of traveling with small children, nor of having too many mouths to feed. On the contrary, the more children, the more hands available to cultivate the land. If there are too many people in one area, some can move to new sites and farm there. When the agricultural revolution began, the possibilities for migration were unlimited: there was an entire planet to occupy.

This expansion in search of new land to cultivate continued in various regions right into this century. In recent times, when some regions were full to bursting—especially those where agriculture began, such as the Middle East, Europe, and China—some have had to seek out new lands or new occupations in far-off places, often overseas.

The Cultivators' Spread

The spread of agriculture from its initial center must have occurred because the local population swiftly saturated the available resources. Some of the earlier cultivators' children inevitably had to go and look for other land nearby. Initially, this was available almost everywhere, so the populations that developed agriculture in the Middle East were able to spread in all directions, including into Europe. Ammermann and I were able to find sound data for Europe that allowed us to create a map showing the start of agriculture in almost all parts of the continent.

Expansion began about nine thousand years ago, fanning out from an area between Iraq and Turkey. This gradual process eventually took in every corner of the continent, sometimes relatively rapidly, as along the Mediterranean coast and the rivers of central Europe, and sometimes more slowly. Having reached the far north, Scandinavia for example, the movement stopped: in those times the climate was still very cold and there were no methods of cold-weather farming. Migrants penetrated farther north only later and more slowly. Agriculture took four thousand

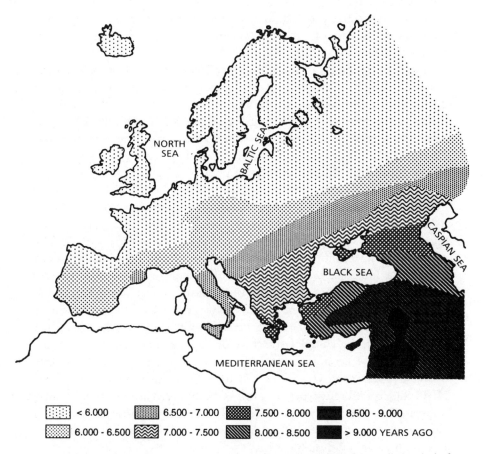

< 6.000	6.500 - 7.000	7.500 - 8.000	8.500 - 9.000
6.000 - 6.500	7.000 - 7.500	8.000 - 8.500	> 9.000 YEARS AGO

6.5 Spread of agriculture in Europe, based on radiocarbon dating of the earliest arrival of Neolithic cultivators in various regions. Map based on research by A. Ammermann and L. L. Cavalli-Sforza.

years to reach the regions farthest from its base (Britain, Denmark, and Spain), about four thousand kilometers away. This means that it traveled at an average rate of about one kilometer a year, as the crow flies.

A grasp of the use of boats, and their advantage as a faster means of expansion than land travel, undoubtedly numbered among the cultivators' skills. A Neolithic boat used on the Seine has recently been found near Paris, but the observed spread of obsidian tools had already revealed that the Neolithic must have known about navigation. Obsidian is a rare, volcanic stone that the cultivators procured en route from Turkey into Greece, first from the islands in the Aegean and later, having reached southern Italy, from the Lipari Islands, where large lodes still exist today.

Obsidian sometimes traveled far from its place of origin, by way of the "first trade," a sort of exchange system that operated solely between adjacent villages.

The spread of early farming took place along the coast of the Mediterranean, and also up the river system of the mainland. Agriculture moved up the Danube's main course and that of its tributaries to their sources, which are close to that of the northward-flowing Rhine, the Elbe, and other waterways. This way the newcomers occupied the whole central European plain. Then, multiplying and spreading repeatedly, they gradually penetrated the farthest corners of the continent.

Throughout this period of expansion, Europe was already inhabited by the peoples who had moved in during the previous thirty to forty thousands years, at the time of modern humans' first great exodus. The hunter-gatherers living in Europe at the time of the arrival of farmers are known as Mesolithics (Middle Stone Age people), and already lived there in the millennia before the arrival of agriculture. Their stone tooling techniques were more advanced than those of the later Paleolithic groups, but they didn't know how to work the land. They may have raised some local crops, but the economy was not as elaborate when compared with the complex structure of cereal and animal farming characteristic of the economies deriving from the Middle East. Mesolithic people not only lived very differently from Neolithic farmers, but probably occupied different types of terrain; the hunter-gatherer sought areas populated by animals, which meant woodlands in particular (most of Europe was forested at the time), while the Neolithic needed to clear the forests to create cultivatable areas.

Destruction of the forests took place slowly, making it feasible for the hunter-gatherers and farmers to survive side by side even in areas where Mesolithic groups were initially more numerous. The most densely populated part of Europe in pre-agricultural times, southwest France and northern Spain, had been the home of the great Magdalenian culture, which was preceded and followed by others going by different names. At the center of this area Basque is spoken today, and there are valid reasons to think that their language descends from one spoken in ancient times by the last Mesolithic hunters inhabiting the region before the cultivators came (see Figure 9.2).

Ammermann and I put forward the theory that the expansion of agriculture was that of the people practicing it, that is the farmers themselves, under the push of a population excess determined by their tendency to have numerous children, more than the hunter-gatherers. It is actually immaterial if more children were born to the farmers, or they

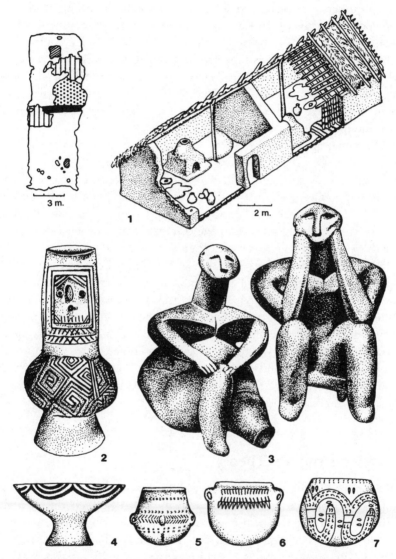

6.6 Neolithic expansion in Europe: (1) floor plan and reconstruction of a Neolithic home; (2) recipient with pedestal from Hungary, 4500 B.C.; (3) terra cotta figurines of man and woman, fourth millennium B.C.; (4–7) clay vases: Cris pot from Romania, Danubian pot, Cardial pot from southern France, and clay pot with linear decoration from Germany.

died less, or both things were true, as long as there was net growth. The important thing is that hunter-gatherers in the area had customs keeping their numbers stationary (more or less constant in time), at the low population density which they had kept from time immemorial. This is usually true today of the few hunters-gatherers left. By contrast, the farmers' population was assumed to be growing in numbers at some rate and would therefore, sooner or later, saturate its immediate surroundings. Farmers would tend to expand to settle in near unoccupied areas, where their cycle of multiplication and migration could start again and continue until there was new land to occupy.

With time, agricultural innovations were introduced to improve land yield and increase the number of people it could support. This happened fairly quickly in the Middle East, where technological progress made possible the emergence of the first urban civilization in human history. City life, however, also led most probably to an increase in the death rate, so demographic growth in the original centers of agriculture must then have slowed back down to a standstill.

During the period of urbanization of the original centers, the outerlying agricultural regions still used very simple farming techniques that have left almost identical archaeological traces over wide areas. Culture in Hungary, Austria, Germany, and France, for example, is remarkably similar and remains unchanged throughout the first millennia of the agricultural era: everywhere there are clay objects decorated in the so-called linear style and similar wood-frame houses covered with dry mud. There simply was no great stimulus to improvement as long as there was essentially virgin land to occupy.

An Unorthodox Theory

Our theory contradicted the prevailing views, particularly in Britain and the United States, where archaeologists had rejected the theory of expansion and migration in favor of the idea that population composition changed very little, and that only ideas and artifacts circulated.

In the interwar period, the British school of archaeologists was persuaded by their brilliant colleague Gordon Childe that each appearance of a type of artifact or innovation (such as axeheads or special sword shapes, cups, or chalices) and its spread into a wider area was accompanied by a movement of people. Nothing was known of these prehistoric expansions, and the hypothetical conquerors who brought their artifacts with them were given the names of the archaeological remains found (for example, Bell beaker or Battle-ax people). It was certainly a mistake to identify these people with the spread of objects. There may have been spreads of customs, fashions, or techniques that were perhaps short term and geographically limited, or territorial gains by an aristocracy leading to the spread not only of customs and objects but also of people coming from a numerically smaller social class, which would qualify not as the spread of a people but of a fraction of a people.

After the war, the English-speaking archaeological world, certainly the most active and prestigious one, had adopted an essentially unanimous rejection of "migrationism," the previous enthusiastic and somewhat uncritical acceptance of all change in artifacts, as evidence of peoples' migrations. The doctrine that replaced it, some times called "indigenism" maintained that it was ideas, and artifacts that circulated. Rarely, if ever, persons changed residence; it was always the indigenous people who accepted new cultures, imitating them from neighbors. In retrospect, this reaction was justifiable for some situations, but misplaced for others. It does show, however, that archaeology can rarely differentiate between the two phenomena, movement of people, or of ideas (in general, physical persons as opposed to their culture).

We called the two explanations *demic*, implying the spread of populations (ours—a very special type of migrationist hypothesis, in which a technological innovation determines a population explosion followed by migration, and hence expansion), and *cultural*, the classical indigenist one (implying the transfer of ideas, technologies, and artifacts). Even if it could not discriminate between the two hypotheses, examination of archaeological data indicated that the expansion of neolithic farmers was very slow but regular. Their rate of expansion was qualitatively and even quantitatively in agreement with demographic calculations based on population growth and migration. An essential demographic factor controlling the rate of expansion was the magnitude of the increase in population density permitted by the arrival of agriculture. This was expected to be large under the demic hypothesis. In fact, the population density of Neolithic farmers was probably ten to fifty times higher than that of the last hunters.

How can archaeological remains be used to judge population density? Generally, the number of settlements (archaeological sites) is counted and the inhabitants assessed on the basis of the number and size of huts. Parallel ethnographic situations can be very useful to this end. Multiplying the number of people and sites provides an idea of density. Naturally, there are various possible sources of error, the most obvious of which is that not all the sites existing have yet been discovered. Also, calculations can take into account only sites analyzed in full, where we feel we know everything it is important to know. In most areas, such exhaustive research either is impossible or has never been undertaken.

The Mesolithic inhabitants of Britain hunted deer, and the number of bones found near their settlements has allowed us to estimate the size of the preagricultural population of the whole island at about five to ten thousand people. This is a tiny figure—the level is now some ten thousand times higher. A historical parallel has been used to check whether this is a reasonable estimate. An approximate count has been made of a population that still lived by hunting and fishing in the 1800s on the island of Tasmania, which is about a third the size of Britain and has a similar climate. When the white settlers arrived, there were about two to three thousand natives in all.

Unfortunately, the entire indigenous population of Tasmania has been eliminated by a combination of the colonials' wish to remove an inconvenient feature of the landscape and by their Western diseases. In Tasmania, therefore, as in so many other places, voluntary and involuntary factors joined forces to wipe out the native population after the arrival of white settlers.

Why did Agriculture Begin?

Leaving aside for the moment the effective validity of our theory, it is sensible to ask why agriculture developed, and why it developed where it did and at a particular moment in history.

The population density in certain zones probably eventually exceeded the limit for survival in the old hunting and gathering economy. This overpopulation went hand in hand with changes in the earth's environment. During the same period, the planetary climate cooled considerably and flora and fauna altered. In America, for example, around eleven thousand years ago the mammoths died out, either because their plant food sources disappeared or because they were hunted to extinction. On the North American plains, the bison took over and became the main food source. Elsewhere, the changes were neither so rapid nor so painless, and human communities must have faced great difficulties.

These two factors explain why agriculture began during much the same period in more than one place, probably where particularly fertile surroundings or the ready availability of plants and animals encouraged higher population density.

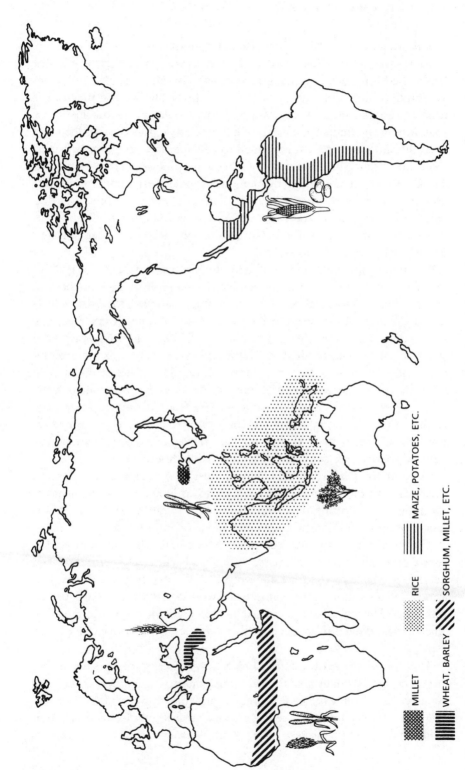

MILLET

RICE

MAIZE, POTATOES, ETC.

WHEAT, BARLEY

SORGHUM, MILLET, ETC.

6.7 The oldest and most important areas of agricultural activity.

This initially occurred in three distinct regions.

One is the Middle East where local wild cereals, particularly wheat and barley, had long been consumed. In today's Israel, a population known as Natufian built stone house settlements, probably because they didn't need to search far for their food and thus could afford the luxury of abandoning the hunter-gatherers' nomadic lifestyle. Once established in a fixed dwelling, they probably wanted to cultivate the terrain in their even more immediate surroundings, thus becoming inevitably fixed on cultivation. In Israel are some of the earliest and best-studied Neolithic examples, but by no means the only ones. There were others farther afield—for example, in Iran, where sites have yielded many other sources of cereals, and some of the earliest traces of sheep and goat farming, dating back about 10,700 years.

The same happened, under different conditions, in China. Agriculture first developed in the north about nine thousand years ago, in the area around the old capital of Xian, where sizable, complete Neolithic millet-farming villages have recently been excavated. Women were especially honored in these areas, to the extent that Chinese archaeologists have suggested they were the first to invent agriculture. This is an interesting suggestion, because in hunter-gatherer communities it is usually the men who hunt and the women who gather. As a result, the women knew more about plant life and had more to gain from setting up fields close to home. It is thought that women were particularly esteemed in northern China, because their tombs are richer. Elsewhere, male burials generally are richer, or there is no identifiable difference.

In southern China, however, in at least two regions, one close to Shanghai and the other to Taiwan (which was linked to the mainland at the time), rice farming developed. Initially, the raising of livestock was less important, despite the abundance of swine in both areas.

The third important area comprises Mexico and the northern Andes, where pumpkin, beans, and maize were already cultivated eight thousand years ago. Maize reached Europe only after the discovery of America. It was originally a small plant whose cobs were only two to three centimeters long. The cobs have increased in size regularly right up to today, thanks to careful cultivation and, probably, the continual selection of the best.

The contribution of Central and South America to agriculture has been enormous: many plants, including potatoes, tomatoes, cocoa, and manioc, have been exported to other continents only recently.

Manioc (also known as cassava, tapioca) grows only in tropical conditions, and is one of the few plants to grow easily. Since it was taken to

1. 5200 – 3400 B.C.

2. 3400 – 2300 B.C.

3. 3300 B.C.– A.D. 700

4 AND 5. 700 – A.D. 1540

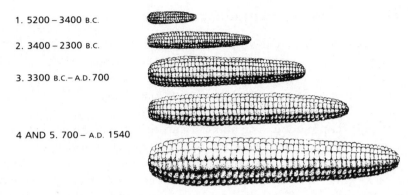

6.8 Growth in the size of the maize cob in Mexico since the start of cultivation. (Actual sizes.)

Africa, probably by missionaries, two or three centuries ago, it has replaced almost all the previous crops in the continent's humid tropical regions, the most important of which had probably been sorghum (Indian millet).

Having reached North Africa, probably from the center in the Middle East, agriculture began spreading to tropical areas. About four thousand years ago, the African farmers who lived in the area of the Sahara, where they owned large herds of cattle, started to abandon the increasingly arid region, which eventually turned into desert. Some of those who emigrated south remained pastoralists, others had to learn how to cultivate numerous new, mostly local, tropical plants.

From the Middle East, agriculture spread in every possible direction, not just toward Europe and northern Africa but also northward into the Steppes and eastward into Iran, Pakistan, and then India. Neolithic farmers proved able geneticists by domesticating many plants and selecting new varieties.

The other centers, too, were sources of major expansion. From China, agriculture spread to Korea, Japan, Tibet, and Southeast Asia; a second route took it from Taiwan to Indonesia, across the Philippines and into Polynesia in the west and Madagascar in the east. Local crop raising sprang up early and flourished in New Guinea, spreading to the nearby Melanesian islands. This allowed a considerable increase in population density, which, unusually, was not accompanied by the development of metal technology. I know that some tribes in the New Guinea interior still made and used stone tools as recently as twenty-five years ago. Agriculture did not reach Australia at all until the white settlers came at the end of the eighteenth century.

Did Humans Spread, or Technology?

I was convinced that the spread of agriculture was accompanied by movements of farmers by my own observations of contemporary African hunter-gatherers (Pygmies), living near Bantu farmers. An important reason is that it is very hard to change one's way of life, and the differences between the hunter-gatherers' lifestyle and the cultivators' were, and are, profound. The hunter wants to remain a hunter-gatherer because, under the right conditions, the lifestyle offers a pleasurable, easy existence. The invention of agriculture was probably dictated by pure necessity: either overpopulation or climatic changes causing fauna and flora changes meant that hunting and gathering no longer granted local human communities adequate survival levels.

At the close of our initial survey, it was clear to Ammermann and me that archaeology alone would not solve the questions raised by our theory. We explored the issue in a book, *Neolithic Transition and the Genetics of European Populations* (Princeton University Press, 1984), which also examines proofs that came from other fields of subsequent genetic surveys. Archaeology had provided some important leads, however, such as the fact that agriculture spread throughout Europe from the Middle East at a nearly constant, slow rate, and caused substantial increases in numbers of inhabitants.

Initially, our theory that the spread of agriculture was caused by movements of farmers, under pressure from demographic density, labored for acceptance in an environment dominated by the opposing view. We had to provide strong supporting evidence. My hope was that genetics could help. But how?

The Contribution of Genetics

Genetic studies had already indicated that there were peoples different from those in and around Asia in the far west of Europe. The idea had been fielded that the Basques descended from the modern humans who occupied Europe (exemplified by Cro-Magnon). This idea was initially prompted by the much higher frequency of Rh− genes found among the Basques. Subsequent analysis brought to light other genetic differences. The Rh− gene map showed that the level of Rh− was much lower in eastern Europe and Asia, and sometimes actually falls to nil the farther outside Europe you go. This suggested that Rh+ was the prevalent type among Neolithics when they began their trek from

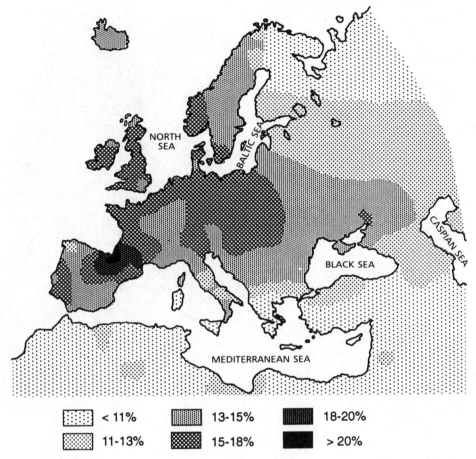

6.9 Genetic frequency maps for Rh−individuals (above) and B genes (page 146) in Europe.

the Middle East, and that, at the other end of the scale, were the largely, or perhaps even exclusively, Rh− indigenous Europeans.

The rhesus gene map, therefore, defended the idea of a Middle Eastern expansion toward Europe of groups with Rh+ blood, which mixed along the way with other peoples who had largely or exclusively Rh− blood. One gene, however, is not enough to substantiate a theory; we needed other corroborating evidence. Migration involves all genes, not just one. It was essential for as many genes as possible to reinforce the same position.

Together with two Italian colleagues who spent some years at Stanford in my laboratory, Paolo Menozzi from Parma University, and Alberto

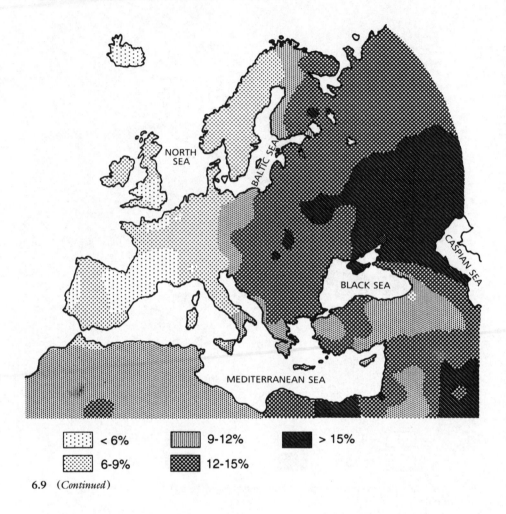

▦	< 6%	▦	9-12%	■	> 15%
▦	6-9%	▦	12-15%		

6.9 (*Continued*)

Piazza from Turin University, we began to feed into the computer all the scientific data available on known blood groups, on HLA genes, which are used to predict organ acceptance for transplants, and numerous other genes that had been isolated and studied in various groups in the meantime.

We drew up and compared maps for numerous genes. Some followed a pattern similar to Rh, and, like Rh+, were especially frequent in easterly areas and especially rare in the west. Occasionally, an absolute maximum (or minimum) peak centered on agriculture's original core in the Middle East. This meant there was a strong chance that genetic maps would confirm our theory. Some genes, however, behaved very differently; for example, the map of the B blood group showed the highest levels to be in southern Russia. It was clearly essential to look at large numbers, construct a common unifying framework, and explore other possible explanations.

Genetic Landscapes

Dealing with large amounts of data and seeking to extrapolate general behavior, as in our survey of thirty-nine different genes in numerous European populations, requires the help of statistics.

Back in the 1930s, an excellent American mathematician, Harold Hotelling, developed a system for summarizing large and complex bodies of data such as our own. This was exploited very little initially, because it involved enormously long calculations. We would have needed an army of mathematicians prepared to carry out myriad arithmetic operations by hand. A few crystallographers, physicists, and the occasional mathematician were the only ones able to organize the large teams of patient, precise individuals willing to undertake such an arduous job. In fact, before modern data processing made calculation easy, fast, and accurate, virtually no one dared apply Hotelling's method.

This procedure, *principal components analysis*, is practically impossible to describe without introducing mathematical concepts unknown to many readers. It allows us to summarize large quantities of data—in our case, all the information on genetic distribution data of all individuals—and discover the trends and patterns common to many genes that are the outcome of events influencing their geographic distribution. The procedure also allows us to isolate and describe the latent "structures" within data, which are probably dictated by historical and geographic factors. In our case these latent structures could be expressed in geographical terms, which could help to understand the nature of the phenomena responsible for them. Individual genes also would show the same phenomena but the picture they provide is hazy and incomplete, because each gene frequency is subject to random oscillations. As with all statistical methods, this problem is overcome by calculating averages from a large number of observations (in this case, data on thirty-nine gene types in many European populations). Mathematicians and physicists not familiar with the method we used may understand its nature noting that the steps include: 1) calculating a geographic map of the gene frequencies of many genes; 2) applying spectral analysis to the values of all gene frequencies at a selected set of nodes; and 3) using eigenvectors corresponding to the highest eigenvalue to generate "synthetic" geographic maps.

The map for each gene (Figure 6.9 uses the example of Rh −) is made of isogenic curves that link areas with the same gene frequency. When an average is taken of many genes, as happens with principal components analysis, the results of all individual genes are substituted in each point of the map by a single number, reflecting the whole group of genes under consideration. This number is the value of the principal component at that point of the "synthetic" map, and we treat it like altitude on a normal geographic map. The outcome is a kind of "genetic landscape" with mountains and valleys, which reveal a latent structure within the data. The next task is to explain the significance of the contour patterns.

This is by no means all that principal components analysis can tell us. The first principal component expresses the highest fraction of genetic variation that can be extracted from the data. Much else also can be gleaned; the map, in fact, has explained only a certain percentage of the variation among the group of genes. For example, in Europe, the first component explains 28 percent of the point-to-point variation in gene frequency; 72 percent of the information contained is still hidden.

At this point we repeated our operations in much the same way to extract a second principal component from the residual variation. This second principal component offers a different, independent genetic landscape—one less important than the first. The second principal component for Europe extracts 22 percent of the initial variation. Repeating the operation again, the third and fourth components can be obtained, and so on up to the number of genes used minus one. The component landscapes obtained are progressively less significant, and after the fifth and sixth their reliability is limited. The most valuable information on the range of genetic variations within a given region has already been extracted.

Menozzi, Piazza, and I applied this method to data on Europe for the first time in 1978. To our considerable satisfaction, we discovered that the highest point of the first principal component centered right on the Middle East, and then diminished with distance from that spot. The very lowest points, in fact, were found in regions like the Basques country, and in general, at the furthest distance from the Middle East. Figure 6.10 expresses the "altitude" of the first principal component. The map clearly suggests a genetic expansion originating in the Middle East, which went on to influence the whole of Europe in a fairly regular way. This map is based on more recent data than the 1978 one and reflects

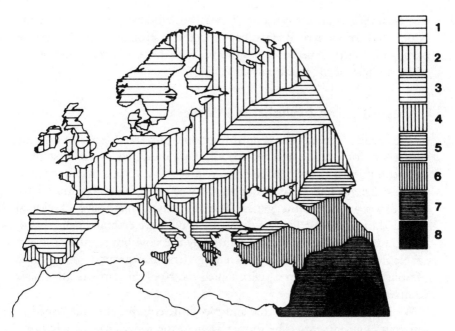

6.10   The most important genetic landscape for Europe (the first principal component of the frequencies of 95 genes). It faithfully reflects the spread of agriculture in Neolithic times (see Figure 6.5), and this is its most probable underlying cause. (The scale of 1 to 8 is arbitrary.)

information collected from a larger number of genes (ninety-five). Despite the considerably greater amount of material at hand, the new map differs very little from its predecessor.

Why, though, should the first principal component of European genes show this specific distribution? When agriculture began, the peoples of Europe were undoubtedly genetically different from each other. For our purposes we include the Middle East in Europe. The genetic variation was especially large at that time because, before the adoption of agriculture, population density was decidedly low and drift very effective. The inhabitants of Europe and the Middle East may have numbered perhaps only one hundred thousand (whereas there are now seven hundred million). They were probably divided into small, relatively isolated groups, and as a result genetic drift had determined high variations, somewhat as it does today in isolated mountain villages.

The last Ice Age played its part, too. The cold weather, which peaked eighteen thousand years ago, split the peoples of central Europe into a western section, including the Cro-Magnon (living in southern France and Spain), and an eastern one. It was probably at this point that the populations of western Europe began to differ strongly from the eastern ones and, perhaps because of genetic drift, became prevalently Rh − while the rest of the world remained Rh + .

Mesolithic and Neolithic Peoples

The strong genetic gradient (or gradual increase/decrease in gene frequency) for Europe and the Middle East indicated by the first principal component is very definite and regular. This means that significant initial differences in gene frequency between the two areas must have existed for at least some genes. The differences may have developed mostly during the last Ice Age, but they could have emerged beforehand. The gradient must have been generated by the slow but regular diffusion of Neolithic farmers into a different genetic background, that of the Mesolithic hunter-gatherers, with which the Neolithic farmers must have admixed gradually.

We have said that Mesolithic and Neolithic communities flourished in different environments. The former wanted the forests for their hunting and foraging activities, whereas the latter needed land for cultivation (obtainable by clearing woodlands). In the outermost areas of expansion, for example, in Spain and Denmark, some Mesolithic groups coexisted alongside the Neolithic for a long time, perhaps because their customs were advanced enough to withstand the competition. The two quite certainly had numerous contacts, but they left no definite signs of conflict in central Europe. The cultivators generally lived in villages and dwellings with no particular protection. They constructed some fenced enclosures that they used for livestock. Only some millennia later, and more particularly in the Iron Age, do we start to find structures erected with defense clearly in mind.

The division of territory between Mesolithic and Neolithic peoples may have allowed a peaceful coexistence and may even have encouraged the exchange of goods and intermarriage. There are always social contacts between hunter-gatherers and neighboring groups of cultivators; we have seen how the African pygmies have maintained relations with the farmers living on the forest edge for thousands of years. There have also been minor demographic exchanges. Despite their conviction that the pygmies are inferior, settled cultivators occasionally marry pygmy women, because in their part of Africa a wife must be bought and pygmy women are more economical. There have been pygmy queens among the tall, proud Tutsi herders of Rwanda. Witchcraft or political advantages, rather than cost considerations, dictated such marriages, as did, some-times, genuine appreciation of the qualities of these women. The reverse marriage—between a cultivator woman and pygmy man—is rare. It is usually acceptable for a woman to move up the social ladder, but not to descend it; in a male-dominated society, the female is much too precious a possession to concede to inferiors. Certain Indian castes practice hypergamy: a woman can marry into a higher caste, but if she marries into a lower one, the couple become virtual outcasts of Hindu society.

There must have been genetic exchanges through marriage because, although there are no longer hunter-gatherers in Europe, their genetic traces can be detected in the continuous gradient of the first principal component. This genetic gradient also tallies almost perfectly with the sequence of archaeological dates for agriculture's appearance in Europe. The first principal component can therefore be explained in terms of Neolithic migration and gradual genetic fusion with the Mesolithic peoples of Europe.

A Computer Simulation

With a number of questions remaining to be resolved, we ran a computer simulation at Stanford with the help of Sabina Rendine, a student of Alberto Piazza's from Turin. A simulation is the simplified computer reconstruction of a phenomenon, in our case the Neolithic migration from the Middle East.

On our computer, we therefore constructed a simplified Europe, inhabited by four hundred tribes of hunter-gatherers. We distributed the groups evenly, because we didn't know exactly where to place them, but we did take into account major physical barriers, such as mountain ranges and, of course, the sea. Simulations are inevitably a rather naive version of reality, unless one is prepared to undertake an exceptionally complex task, which is not generally necessary. For example, we recon-structed the tribes' behavior but not that of individuals, and we did not attempt to accurately simulate the fact that the Neolithic newcomers must have traveled along the Mediterranean coast and European river-ways.

Our simulation applied gradual population growth to each Neolithic tribe until they far outnumbered the hunter-gatherers, on the basis of the cultivators' ability to produce larger quantities of their own food, instead of having to rely on food found in the wild. As Neolithic population density neared saturation point, a progressive falloff to zero growth was applied, but we still generated emigrants who occupied adjacent areas in all directions. Members of neighboring farming tribes in different areas swapped through marriage, as happens almost universally in tribes around the world, and there was a minor influx of hunter-gatherers into local agricultural tribes. The original number of hunter-gatherers was small, in line with archaeological observations, and could not increase in density; their one-way migration to local farming tribes was an imperfect, but practical, means of simulating the switch from a more primitive economy to an agricultural one. In fact, hunter-gatherers disappeared everywhere a few hundred or thousands of years after the arrival of the settled cultivators.

We gave each population a genetic makeup of a moderately large number of genes (twenty). We then simulated the spread throughout Europe of these genes, which we originally attributed to the first farming peoples, in the Middle East. We expected that the analysis of principal components would give us a geographic representation of the migratory flow from the Middle East to Europe generated by the spread of the cultivators. Then we could check the influence of possible complicating factors arising from the end of the cultivators' expansion some five thousand years ago and the subsequent takeover by other immigrants. Small-scale exchanges between villages (which our simulation treated as exchanges between neighboring tribes) undoubtedly took place throughout and beyond the period of Neolithic expansion. There were also, however, larger movements of people, some of which are recognized historically, and there may have been others as well.

We concluded that small-scale migration, comparable in size to that of the last three centuries among the farming communities of northern Italy (which I had looked at, for example, in the Parma valley survey), could obscure only marginally the gradient generated by the spread of Neolithic groups, even if projected over a period of five thousand years.

Simulations of other major migrations similar to the Neolithic one allowed us to demonstrate that principal components analysis often can separate expansions originating in different places. Within the simulation, in fact, each generates a different genetic landscape, indicating the origin and most important area of expansion.

Principal component maps can suggest the existence of prehistoric expansions that are so far unknown to us, but do not help to date movements, unless in relation to reliably datable archaeological facts. That the earliest migrations tend to have the greatest genetic effect can provide a general qualitative idea of the date of a period of expansion, since migrations occur when the population density is still fairly low and genetic drift is therefore more active in determining greater differences in gene frequencies. For this reason the oldest migrations have often influenced the first principal component.

An Unorthodox Idea Proven

This first study of Neolithic expansion in Europe in a genetic light, carried out with genetic data and a previously unused method of analysis, was published in 1978 in the journal *Science*, under the joint signature of Menozzi, Piazza, and myself. To substantiate the theory put forward with Ammermann six years earlier—that the spread of agriculture from the Middle East into Europe was the result of geographic expansions of cultivators—was, naturally, very satisfying. The same article also described the second and third principal components, which suggested other events and phenomena. The second component showed a north-south trend, which we interpreted as the probable result of natural selection due to climate. The third map, equally interesting, showed Poland and the Ukraine as the probable center of a further expansion. We did not attempt to explain this last component.

Our conclusion acquired even greater credibility after confirmation by a group of American researchers headed by Robert Sokal from Stony Brook, Long Island, who used quite different statistical methods. Their results were published in three papers from 1982 to 1991.

Ten years later, we reran the same analysis first tried in 1978, inserting new data published in the meantime. This allowed us to increase the number of genes two and a half times. We could confirm our conclusions and understand the second, third, fourth, and fifth principal components.

The Genetic Dissection of Europe

Concerning Europe, there are strong reasons to consider the first five principal components in terms of expansion. The others are not statistically reliable enough to hazard an interpretation.

At the start of our survey, we had noticed that the second principal component showed a north-south trend, possibly to do with genetic adaptation to climate. Further work has since shown a correlation between these genetic differences and the linguistic differences between the peoples speaking Indo-European and Uralic languages. This in itself pointed to an explanation in terms of population movement and would

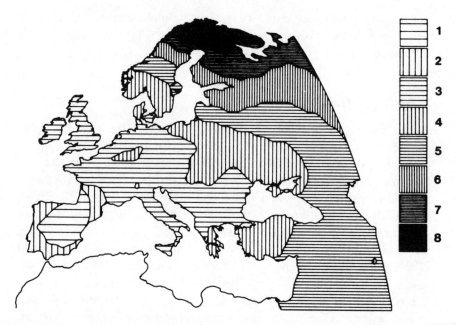

6.11 The second principal component of the genetic map of Europe generates a different landscape. It may represent genetic adaptation to the cold of northern Europe, but it should also be considered in relation to the distribution of the Uralic family of languages (Figure 7.1). Both phenomena are probably the result of a single large-scale migration. The Uralic languages are found mainly in the most northerly regions of Europe and western Asia, and are presumably a legacy of the groups who pushed north in ancient times and gradually adapted to arctic conditions. (The scale of 1 to 8 is arbitrary.)

not contradict the initial interpretation, because the speakers of Uralic languages have traditionally inhabited very cold regions.

The peoples of Siberia, who are physically similar to Mongols, have developed special resistance to the cold, which makes them genetically a bit different from the populations living farther south. In the hundreds or thousands of years of separation from other ethnic groups, a new linguistic family, Uralic, emerged in western Siberia. In northern Russia and southern Scandinavia, other European-type peoples speaking different languages arrived and partially merged with the Uralic ones.

The second component (Fig. 6.11) therefore shows a north-south gradient that reflects diversification in climate and linguistic type. The most typical representatives of the Siberian physical type and Uralic language are found east of the Ural mountains, in Asia. In Europe, however, we find languages of a Uralic group, known as Ugro-Finnic, which includes Finnish, Hungarian, some Baltic languages, and various

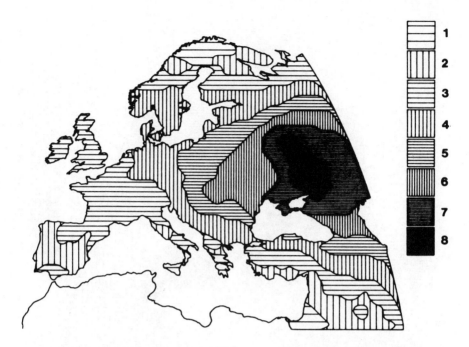

6.12 The third principal component of European genes strongly correlates to the map of archaeological data (see Figure 6.15), which—according to M. Gimbutas's interpretation—reflects the spread of Indo-European language-speaking nomadic herders from the Euro-Asiatic Steppes, between 4,500 and 6,000 years ago. These were probably descendants of the first cultivators migrating to the Steppes north of the area in which agriculture originated. Horse-rearing—common in these regions—was a local adaptation. (The scale of 1 to 8 is arbitrary.)

Lapp ones spoken in northern Scandinavia. In physical terms, the Lapps show the greatest marks of Siberian origin, but there are faint traces of genetic influence among the Hungarians and Finn as well.

The third component (Fig. 6.12) revealed strong links with an event distinct from the spread of agriculture: a secondary expansion, related to the development of animal grazing in southern Russia. We will return to this when examining Indo-European languages.

The fourth component (Fig. 6.13) strongly resembles the Greek expansion, which reached its peak in historical times, around 1000 and 500 B.C., but which certainly began earlier.

The fifth map (Fig. 6.14) again concerns agriculture, but in a negative sense, since it shows the genetic makeup of the Upper Paleolithic and Mesolithic peoples of western Europe, who at least partially withstood the expansion of the cultivators. They managed to survive without being

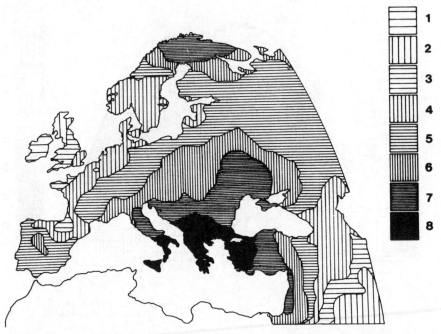

6.13 Fourth principal component for Europe, which very probably reflects the Greek expansion of the second and first millennia B.C. (The scale of 1 to 8 is arbitrary.)

fully absorbed and therefore remained somewhat genetically distinct from their neighbors. The dark area of the map corresponds to the region in which the language that became modern Basque was once spoken. Today, in France at least, the Basque-speaking area is smaller than that shown on the map. However, place names on the French side of the Pyrenees betray the origins shared with the oldest Basque regions, and it is well known that place names may survive for millenia.

Comparison of this component with other maps could give the impression that the Basque area was a center of expansion. Principal components, however, do not reveal whether we should speak of an explosion or an implosion—whether, that is, today's residents of the Basque-speaking areas expanded or if they were concentrated inward under pressure from external migrations. On the basis of what we know, it is much more probable that this is a residual population that has resisted genetic, linguistic, and cultural infiltration from neighbors. We are tempted to coin a new word, *impansion*. Naturally, there is farming in today's Basque regions, but the newspapers show that the Basques' desire

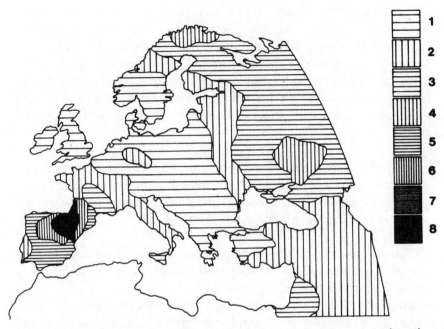

6.14 Fifth principal component of European genes. The dark area corresponds to that where Basque is still spoken or was until a few centuries ago. It also corresponds to areas where Basque place names and artistic styles are common (Figure 11.2). Here the dark area indicates not an expansion (as is probably the case for the first, second, third, and fourth principal components) but the influence of residual pre-Neolithic peoples who have not been completely absorbed by the Neolithic groups expanding westward.

to resist aggression from outside cultures and affirm their own autonomy has by no means diminished over the years.

Multiplication and Migration: Factors of Expansion ("Diasporas")

Extending our genetic landscape analyses beyond Europe, we realized that there are significant traces of numerous phases of expansion around the world, enough to convince us that the history of modern humans has been punctuated by repeated migrations with something in common: a technological advantage that could be handed on to children and that allowed high enough population growth to provoke sustained emigration levels. The word *migration* has more than one meaning; it can describe the simple movement of a people, but an expansion is a centrifugal migration stimulated by local demographic growth.

The introduction of agriculture is perhaps the most powerful example of this, because over the last ten thousand years it has allowed a thousandfold or more increase in population, from millions to billions. The expansion of modern humans around the world from one hundred thousand (or perhaps we should say fifty to sixty thousand) years ago to ten thousand years ago determined perhaps one hundredfold growth. All these dates are uncertain, but we can pinpoint some of the fundamental stages:

- The first modern humans lived about one hundred thousand years ago, perhaps numbering between twenty and one hundred thousand people (estimates based on uncertain data). They inhabited the areas where the modern human race first developed—eastern Africa or the Middle East (or both).
- Their expansion ended about ten to fifteen thousand years ago. Having reached virtually all the parts of the globe inhabited today, they numbered about five million.
- Livestock and crop-raising activities began nine to ten thousand years ago. The inadequate food supply provided by a purely hunting and gathering economy, at least in temperate regions, stimulated the start of food production in more than one area, generating an unprecedented increase in population density.
- Certain animals provided special opportunities for expansion. The horse, used for food, for transport, and as an instrument of war, was also responsible, from five thousand years ago, for the spread of nomadic herders from southern Russia into Europe, central Asia, and India. Domesticated camels helped the Arabs expand into northern Africa in Christian times. In the southern Andes, the llama was one source of wealth of the Inca empire and was used to transport food and other goods.
- Transport was facilitated by the domestication of animals but also by numerous inventions, such as the wheel, the sail, the outrigger canoe for ocean navigation, and the compass, as well as by the study of the position and course of the stars.
- Military innovations facilitated expansion through conquest; in addition to the use of the horse, these included metal defense and attack weapons, first in bronze and later in iron.

Among these examples, an innovative factor often plays a predominant or essential role in each stage. We don't yet know which factor determined the first phase (the initial spread of modern humans across the face of the earth), but we can suggest some possibilities:

- *The development of advanced language, which permitted better communication between individuals and groups and thus facilitated expansion to totally new areas.* The splintering of prehistoric cultures in the period of modern human expansion noted by Glynn Isaac took place in the last fifty thousand years. The names for different cultures proliferated rapidly. Distinct languages and dialects were probably developing along with the diversification of stone and bone crafts of different places. If this cultural variety shared common roots with the diversification of languages and dialects that separate ethnic groups it could facilitate not only cultural, but perhaps also genetic diversification. If, as seems likely, human language has made huge leaps forward in the last fifty to one hundred thousand years, some sort of biological evolution must have permitted it. Therefore, the innovations were not only cultural and technological, but also biological.
- *Improvements in transport were probably essential for travel to distant regions.* Australia could be reached only by undertaking an arduous crossing of as much as forty to fifty miles of sea. No remains of the boats, rafts, or other vessels used for these crossings have been found. Simple tree trunks may have been used (although crossing forty to fifty miles of water clinging to a tree trunk seems an unlikely feat). In any case, their vessels must have been made of wood, and wood rarely comes down to us intact.
- *Expansion to areas with profoundly different climates, which led to significant biological and cultural adaptation.* The latter finds expression in new ways to build dwellings and make clothes as well as in improved hunting and fishing techniques.

The Other Great Migrations

Today we can reconstruct the probable causes of some expansions that have left identifiable genetic traces. For others, we must rely on archaeological data. Still others, which occurred in areas where archaeological research has been limited or nil, stimulate the search for vanished civilizations. In many areas of the world, not enough archaeological data exist to create reliable genetic landscapes. For now, we can hypothesize from genetic maps a number of probable expansions, suggested in part by linguistic evidence.

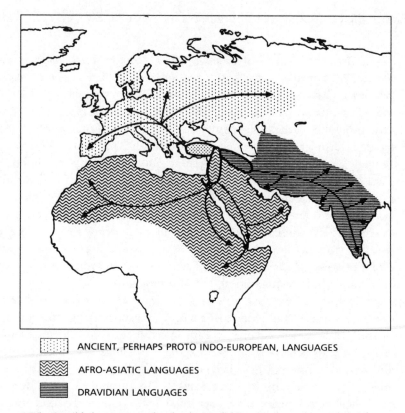

ANCIENT, PERHAPS PROTO INDO-EUROPEAN, LANGUAGES

AFRO-ASIATIC LANGUAGES

DRAVIDIAN LANGUAGES

6.15 Possible languages spoken by the Neolithic cultivators who expanded from the Middle East, independently proposed by L. Luca Cavalli-Sforza and Colin Renfrew.

The routes from the original Middle Eastern center of agriculture lead not only toward Europe but also in the opposite direction, toward Iran, Pakistan, and India. We know of a farming civilization in Pakistan, the Indus Valley civilization, that reached its peak between thirty-five and forty-five hundred years ago (later than the Euphrates and Tigris civilizations of the Middle East). Two cities, Moenjo Daro and Harappa, grew especially large; each numbered about fifty thousand inhabitants at its zenith. They were abandoned about thirty-five hundred years ago, probably after a change in the course of the Indus River, and were never rebuilt. At that time Asiatic nomadic herders arrived in the area, bringing Indo-European languages to India, Pakistan, and Iran.

Several different expansions have clearly taken place in Africa, one in North Africa, coming from the Middle East, which is partially explained

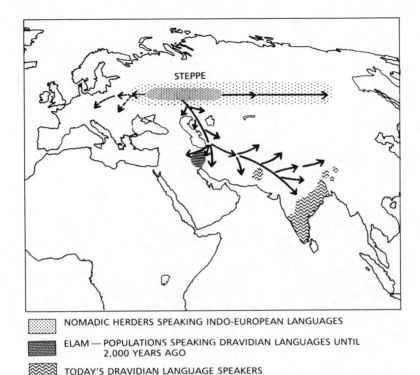

▒▒▒ NOMADIC HERDERS SPEAKING INDO-EUROPEAN LANGUAGES

▬▬▬ ELAM — POPULATIONS SPEAKING DRAVIDIAN LANGUAGES UNTIL
2,000 YEARS AGO

〰〰 TODAY'S DRAVIDIAN LANGUAGE SPEAKERS

6.16 Expansion of the Steppe nomads, who spoke Indo-European languages, into Iran and India and perhaps also into Europe (dotted line). These are the so-called Aryans, a term meaning "noblemen." They may originally have been related to the populations indicated in the third principal component map (Figure 6.12), who spread into Europe. They arrived in India about 3,500 years ago. In Iran and India the nomads' Indo-European languages have largely replaced the earlier Dravidian languages.

as the first spread of agriculture. It was followed by the expansion of Bantu-speaking peoples, who originated between Nigeria and Cameroon and spread eastward but above all southward, between five hundred and thirty-five hundred years ago. Another equally recent migration started somewhere between southern Arabia and Ethiopia and developed in both directions. An Arabo-Ethiopian kingdom commenced about three thousand years ago in southern Arabia, with its capital at Saba (home of the Queen of Sheba). Subsequently, the capital moved to Axum, in northern Ethiopia. Earlier migrations from Ethiopia may have occurred. There were very probably agricultural expansions in western Africa before the Bantu one, but we lack precise archaeological evidence of these.

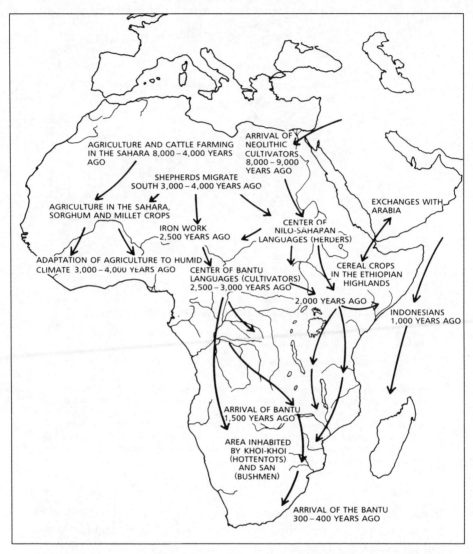

6.17   Probable recent expansions within Africa, based on genetic landscapes. At least one, the latest, known as the Bantu expansion, tallies well with archaeological data and with the linguistic data that first suggested the idea.

In China, expansion from the first millet-producing agricultural areas was limited in the west and north by desert and steppes. To the south, the climatic conditions favored development of at least two agricultural civilizations based mainly on rice, which began slightly later. There are profound genetic differences between the northern and southern Chinese, which probably reflect an ancient diversification caused by the development of two separate agricultural cultures in the north and south; the eastern region, around Shanghai, is in some ways intermediate both genetically and culturally, although it also shows signs of independent evolution.

The second principal component of Asia suggests an expansion from the area around the Sea of Japan, centered on Japan and Korea. Between ten and fifteen thousand years ago, Japan was linked to the mainland—with Russia in the north and Korea in the south—and had its own internal sea. As a result, the area developed uniquely on the back of the sea's fishing potential. The earliest domestic pottery originated in Japan about eleven thousand years ago. The population five to six thousand years ago was already very large and numbered some three hundred thousand, thanks to the development of fishing techniques and the abundance of wild plants. Later the population level declined a little. These developments predate agriculture, which came very late to Japan (little more than two thousand years ago) from Korea. As mentioned, principal component maps don't help determine migration dates, except for the earliest migrations, which tend to leave the largest traces. On this basis, it is possible that the Japanese expansion may be linked to one of the Paleolithic migrations from eastern Asia to America.

Improvements in the quality and quantity of genetic and archaeological data can, with judicious joint use, open up new avenues in our understanding of the earliest human history.

SEVEN

The Tower of Babel

The remains of Babylon lie fifty miles south of Baghdad. In the language of the day (Akkadian), the city was known as Babilani, meaning "Gateway of the Gods." In the grounds of the temple to Marduk, the city's most important god, stood the Etemenanki, or "House of the foundation of Sky and Earth," known to posterity as the Tower of Babel. It was a seven-story stepped pyramid construction, a three-hundred foot "ziggurat." Herodotus describes it in detail.

The Bible says that the Babylonians wished to build a mighty city with a tower that reached the sky. God, who was not pleased about this ambitious project, decided to prevent it by having all the construction workers speak different languages; in the ensuing confusion, building became impossible. In Hebrew, *balal* means to confuse, and the Bible plays on the similarity: "Therefore is the name of it called Babel; because the Lord did there confound the language of all the earth; and from thence did the Lord scatter them abroad upon the face of all the Earth" (Genesis 11:9). This is the biblical version of how human linguistic differences came about.

Thoughts About a Legend

The differences between languages are clearly enormous and can easily render cooperation with others difficult. Given their magnitude, the idea that these differences all materialized in the same instant is hard to believe. It is more likely that they developed over a long period of time, and that the problem the Bible alludes to, if there was one, arose because the tower's building site teemed with workers from different regions who couldn't communicate with one another, and consequently quarreled. Language differences must have emerged well before, but Genesis notoriously shortens times by very large factors. The evolution of life, which may have taken three and a half billion years, is reduced to one week.

Languages are defined as different when they are mutually incomprehensible. Small differences that don't hinder comprehension are called "dialectal." In reality, some dialects are practically impossible to understand without translation, which makes us wonder if it is right to exclude them from the category of languages. Historical, geographic, and sociological factors all explain extreme dialectal differences. A trip into rural Italy, England, France, or Spain for example is enough to show just how huge a range of dialects one language can generate.

In certain regions, more than one distinct language is spoken. In the Iberian peninsula, Basque, Catalan, Spanish, and Portuguese all are used. Basque is spoken by about a million people in northeast Spain and also by a few tens of thousands of people north of the Pyrenees in southwest France. Some ethnic minorities in Italy still use their traditional language (French in the northwest, German and Slovenian in the northeast, and Greek and Albanian in some parts of the south). Belgium and Switzerland are both divided linguistically into large sections. In North America, people move around fast, and the time for differentiation of the English language has been short, but there are some differences in accent that frequently distinguish the origin of people.

In addition to variations from area to area, many temporal changes in language are recent enough to be clearly etched into history. That it doesn't take long for one language to become another is also apparent. Until around fifteen hundred years ago, Latin was spoken in western Europe, but the Europeans of today would not be able to converse with their ancestors, except in simple expressions. Separate, mutually incomprehensible languages have emerged in Italy, France, and Spain, all of which, however, continue to manifest their common Latin origins: a geographically more distant cousin is found in Romania, a country whose name alone reveals its past links with the Roman world.

Fifteen hundred years is ample time in which to lose mutual comprehension. For example, Iceland was colonized by the Norwegians at the end of the ninth century A.D. Today's Icelanders, with considerable effort, can understand people from the Scandinavian peninsula, but the Scandinavians hardly understand the Icelanders. A thousand years is the minimum time span for a language to change so much that it becomes incomprehensible.

A Multitude of Tongues and One Language

If we look farther afield in time and space, we come across an astonishing range of linguistic variations. Perhaps the most exceptional are the so-called Khoisan languages, spoken by Bushmen and Hottentots, the South African natives encountered by Dutch colonizers settling in the Cape Town area around 1650. The Khoisan language uses a special range of sounds, produced by clicking the tongue in the way used to call a horse or imitate its trot: there are four or five of these, in addition to the more usual vowels and consonants. This clicking technique is unique to the Khoisan or groups that have had recent contact with them.

Another interesting example is the word *ma* for "mother," one of the very few Chinese words that resemble a common European one. We must quickly add, however, that the Chinese syllable *ma* can be pronounced four different ways corresponding to different musical intonations, and that only one of these means "mother"; the others mean "hemp cloth," "horse," and "to scold," and are all written differently.

Despite large-scale differences, a number of fundamental common denominators link all languages to a single base. Whatever his or her origin, a person is capable of perfecting knowledge of any language, so long as the learning process takes place in childhood. In the early years of life, children have not only an enormous capacity to learn but also a genuine drive (which would once have been defined as "instinct") to learn to speak. A person who doesn't master a language at this early age may never speak it correctly. After puberty, it is very difficult to learn pronunciation of a foreign language well enough to be mistaken for a native. There is not much future in a career in spying therefore, unless you learn the enemy language very young in life.

All the languages in existence are similarly complex in structure; the languages of the poorest aboriginal tribes are every bit as rich as ours and sometimes structurally more complex. They have literature and poetry, even if they are both oral traditions. In fact, the vast majority of languages have never, or only recently, been written down.

How Many Languages Are There?

About five thousand languages, and a far greater number of dialects, are still in use around the world. Many are spoken by only a few hundred people and are destined to die out, just as many have died out in past centuries. Others are in the final stages of extinction or have recently disappeared. Thirty years ago, I met the mayor of Montecarlo, who was one of only four surviving speakers of the local dialect (classified as an Italian dialect from the coast of nearby Liguria in Italy). He wrote a grammar book to save his language from total oblivion. By now, the very last speaker has probably died.

Linguistic diversity can take many forms: sound (phonetics), meaning (semantics), grammar, and syntax can all differ. Equivalents of the Latin word *mater* are *mother* in English, *madre* in Italian and Spanish, *mère* (pronounced "mare") in French, *Mutter* in German, *Mor* in Swedish, *mat* in Russian, and *metéra* in Greek. In all cases the first letter, *m*, has survived, but the second consonant is not always present, and the vowel changes frequently (with his usual sardonic wit, Voltaire commented that consonants are of little help in etymological analysis and vowels none whatsoever). In the ancestral language common to the Indo-European languages, the reconstructed word is *ma*.

How Fast do Languages Change?

If we compare a language with its thousand year-old forebear, we find (an average of) 86 percent cognates (discussed on page 171). If we compare two languages that separated one thousand years ago, both may display an 86 percent residual affinity with the original language. If there have been no exchanges between the resulting languages, and evolution is independent, the reciprocal percent of cognates between the two derivatives will be 86 percent of 86 percent (74 percent), since both should alter from the shared starting point in a similar fashion. The case of Icelandic, which broke away from Norwegian one thousand years ago, is a known exception to the rule: perhaps because of its isolation, it has changed less than other Scandinavian languages.

This criterion gives us hope of eventually reconstructing the story and even the times of the separations of languages. The results of glottochronology are generally approximated, since there are various potential sources of errors. There is, however, a surprising similarity between this method and the molecular clock we spoke of when looking at biological evolution. Both are based on the same hypothesis. Glottochronology works on the assumption that there is a fixed probability of semantic alteration over a given unit of time, leading to expression of a meaning by way of a new word. In molecular evolution a DNA nucleotide is replaced by another, or a protein amino acid by another amino acid, with a fixed probability over time. These probabilities are not as constant as we would like; they sometimes vary significantly between different words. In fact, so does their biological equivalent, but to a lesser extent.

This means that the calculated glottochronological separation times are not very rigorous. Possible correction methods have been suggested, but these are yet to be tested properly.

Who Bites Whom?

For a journalist, a dog who bites a man is not news, but a man who bites a dog is. The order of words is clearly very important, particularly that of the subject (S), verb (V), and object (O). Linguistic change is not limited to semantics and phonetics; both grammar and syntax also can be influenced. Anyone who knows German will be aware, for example, that in many cases, the verb of a subordinate clause is placed at the end of the sentence, "the dog his master bites." In English, French, Spanish, and Italian, the normal order is SVO; in German, it is frequently SOV.

A statistical survey has revealed that in languages the world over, the most frequent orders are SVO and SOV, accounting for about 75 percent of the many languages examined. VSO, with verb first and then the subject, accounts for 10 to 15 percent of cases (and includes Welsh, for example). VOS (verb first, with subject at the end of the phrase) is found in the Malagasy languages, which originated in far-off Indonesia. A number of languages in the Amazon basin would use "the master bites the dog" (OVS) to express the idea of a dog biting its master. There are a few other examples of OSV in America, and the sequence is acceptable in Japanese, too, where the normal usage is, however, SOV. In the same way, numerous other rules for grammar and syntax vary, but on the whole they are more stable than either phonetics or semantics. As a result, the origin of a language is more reliably predicted from its structure than from its phonetics or semantics, even though phonetics is quite impervious to the influence of other languages.

English has borrowed a huge number of Latin-based words (about 50 percent of its entire vocabulary). This is the result of Roman occupation, which introduced Latin; the Norman conquest (with the Battle of Hastings in A.D. 1066), which introduced the use of French; and, finally, the Renaissance. The structure of modern English, however, is that of an Anglo-Saxon language, although a much simplified one.

Contemporary humans tend to classify everything, and naturally enough, attempts have been made to classify languages. A classic example that caused a great stir is the recognition by the British jurist and Orientalist Sir William Jones that the oldest Indian language known, Sanskrit, used for philosophic and religious subjects, undoubtedly shares some features of the ancient Mediterranean languages, Latin and Greek. In a famous speech at the Royal Asiatic Society in Calcutta in 1786, Jones demonstrated that six groups of kindred languages—Sanskrit, Greek, Latin, and probably Gothic, Celtic and Persian—"originated from a common source which most likely no longer exists."

The similarities between Latin-derivatives (such as French, Italian, and Spanish), Germanic languages, and Slav ones had already been observed. Two hundred years earlier, an Italian, Filippo Sassetti, had already noticed the similarity between Sanskrit and Italian. Sassetti lived in India from 1581 to 1588, but it was Jones who first recognized the family known as Indo-European. The nineteenth century saw the start of intense academic interest in linguistics directed for the most part at this family, which is by far the most fully analyzed.

At the start of the twentieth century, the relationship was noted between Indo-European languages and an extinct one used in documents from western China, dated around the seventh century A.D. These revealed the existence of two similar languages, known as Tocharian A and B. In much the same years, clay tablets from the capital of the Hittite empire, which flourished in Turkey between 1500 and 1200 B.C. were deciphered successfully, because they were written in cuneiform characters, whose phonetic meaning we know. This revealed the existence of another Indo-European language, Hittite.

In the 1800s recognition of other linguistic families different from the Indo-European one also began, and this opened new avenues for research, which continued into the 1900s. There is still passionate debate about some of these families because the profoundly different methods adopted by various linguists lead to diverse and conflicting results.

The following table shows the method used by the principal contemporary taxonomist, Joseph Greenberg from Stanford University, which consists of comparing several hundred words in hundreds of languages. The words are taken from those that change least over time, such as the numbers one, two, and three, along with parts of the body, universal aspects of nature, personal pronouns, certain grammatical rules and so on. Many of these are among the first words a child learns, a fact that probably rendered them less liable to change.

Language	One	Two	Three	Head	Eye	Tooth
Irish	aon	dau	tri	ceann	suil	fiacal
Welsh	un	do	tri	pen	ligad	dant
Danish	en	to	tre	hoved	öje	tand
Swedish	en	to	tre	huvud	öga	tand
German	ain	zwai	draj	kopf	auge	zahn
Italian	uno	due	tre	testa	okkjo	dente
Spanish	un	dos	tres	kabesa	oho	diente
French	ön	dö	trwa	tet	öj	dan
Romanian	un	doi	trej	kap	okju	dinte
Albanian	nji	dy	tre	krye	sy	dami
Greek	enas	dyo	tris	kefali	mati	dhondi
Polish	jeden	dva	tsi	glova	oko	zab
Russian	adin	dva	tri	galavá	oko	zup
Finnish	yksi	kaksi	kolme	pää	silme	hammas
Estonian	üks	kaks	kolm	pea	silm	hambaid
Hungarian	egy	ket	harom	fo	sem	fog
Basque	bat	bi	iru	buru	begi	ortz

The table above shows the pronunciation of the words whose English meaning is written in the first row, using a simplified phonetic alphabet. Only the vowels require some explanation:

The vowel "a" should be read as the vowel of the word "not" in American English.

The vowel "e" should be read like the vowel in "bet."

The vowel "i" should be read as the vowel in "me."

The vowel "o" should be read as the vowel in "port."

The vowel "u" should be read as the vowel in "boot."

The reader should not expect this short guide to be perfect. It is difficult to render all the nuances of foreign sounds to the uninitiated reader.

There is little difficulty in believing that words like *tri* in Celtic languages, *tre* in Scandinavian languages, *draj* in German, *trwa* in French, *tre* in Italian, *tres* in Spanish, all meaning "three," have the same origin. It is also easy to believe that the word *kolme* in Finnish, also meaning "three" must have a different origin, and the same is true of *iru*, which has the same meaning in Basque. It may be more difficult to believe that the English "eye" (pronounced *aj*) and Italian "occhio" (pronounced *okkjo*) or German "auge" have also the same origin, but they do. The rules for changes in pronunciation are well known, even if the process of transformation of sounds is sometimes complex. In general there is little room for doubt, and the history of the languages often supply independent confirmation by showing the intermediate steps, in the process of change. Words descending from a common root are called "cognates," meaning "related." English words "water," "wasser" (German), and "vatn" in Icelandic are cognate; but they are not cognate with Ltain "aqua," Italian "acqua," French "eau," which have the same meaning as water, etc., and are cognates among themselves.

Greenberg notes that even a simple table like the one on p. 170 immediately illustrates the systematic grouping of modern Indo-European languages that is now universally recognized.

The first fourteen languages of the table are all Indo-European: two are Celtic, four are Germanic (including English at the top), three Romance (Latin derivatives, that is), two are Slavic, and two (Albanian and Greek) are isolated cases. The last four are not Indo-European; of these, Finnish, Estonian, and Hungarian belong to the so-called Uralic family, which comprises the languages spoken in northern Europe in and around the Ural mountains, which divide Europe from Asia; the last is Basque, which does not belong to any known linguistic family but has some distant affinity with the family of languages spoken in the southern Caucasus. The total lack of similarity between Basque and the other languages in the table is obvious. There are many similarities between the three Uralic languages (although Hungarian stands out more), but there are few (none clearly visible in the table), between the Uralic group and the Indo-European languages.

This table clearly illustrates that there are at least two linguistic families in Europe and one isolated language.

Two American linguists, Morris Swadesh and Robert Lees, have invented a method of evolutionary analysis of languages that has a close resemblance to the analysis of molecular evolution in biology. Called lexical statistics, or "glottochronology," it estimates the percentage of cognatic words among two languages. Swadesh and Lees created a standard list of one hundred words and carried a number of estimations of percentage of cognates among pairs of languages. They discovered that the percentage of cognate words in two languages decreases with considerable regularity, the longer the separation in time of two languages. For instance, the Latin spoken at Augustus' time, two thousand years ago, can be compared with languages that descended from Latin and are spoken today, like Italian, French, Spanish, or Romanian, and the percentage of cognate words is estimated. When this procedure is repeated with pairs of languages that have a shorter or longer known separation time, one can build a curve relating percentage of cognates and time of separation. This curve can then be used to calculate the unknown separation time of two languages, on the basis of the percentage of cognates between them. The list of words prepared by Swadesh and Lee was made using words that change less than average, in order to be able to estimate relatively long times of separation.

One of the most controversial linguistic issues today concerns the Native American language families, which until recently were estimated as at least sixty. According to a recent analysis by Greenberg, there are in fact only three:

1. One formed by the group of nine *Eskimo* languages (known as Eskimo-Aleut) spoken in the far north.
2. One formed by the thirty-four *Na-Dene* languages used mainly in western Canada, but also by two U.S. Indian tribes that broke away from their northern cousins one thousand years ago, the Apache and Navajo.
3. One known as *Amerind*, formed by all the other languages of which there are 583.

This last group, which includes some languages that are extinct but fairly well known, is somewhat heterogeneous but nonetheless recognizably a single unit.

The gap between the view of Greenberg and a handful of his colleagues, and that of other Americanists who work using the completely different approach of binary language comparisons, is enormous. This traditional approach rarely extends analysis to more than two languages, and languages are not classified as related unless they are extraordinarily similar. With 583 languages to consider, there are 169,653 possible combinations of two, and researchers believe that it takes more or less a lifetime of work to decide the degree of relationship between a pair of languages. So far, they have created an enormous number of different families, at least sixty if not, as many think, more than one hundred, where according to Greenberg, there is only one Amerind family (the third mentioned above). The scenario is a little disconcerting if you bear in mind that only fourteen linguistic families exist in the whole rest of the world.

Why is the difference of opinion so enormous? Linguistics is not unique in this sense. In other disciplines, too, such as zoology and botany, there is sometimes an abyss between classifiers who prefer a broad, all-embracing categorization and others who prefer to emphasize detail and differences. We call those who like to group animals or plants —or, in this case, languages—into a few major groups "lumpers," and those who aim to create a multitude of small groups "splitters." In Roget's Thesaurus, the word splitter is synonymous with hair-splitter, quibbler, and casuist. The famous lexicographer apparently wasn't fond of those who reject a synthetic approach to matters and go into detailed descriptions without seeking a broader overview.

The Tireless Classifiers

The debate on the linguistic families of the Amerind is by no means over, but agreement on the other linguistic families is not universal either. The question, then, is "What makes a family?" The word *phylum* is often used instead of *family*. There is a certain similarity between linguistics and biology in this sphere; in biology there have been, and still are, arguments about the divisions of phyla and attribution of organisms or groups of organisms to one or the other, but they are generally less heated. Furthermore, the levels of biological classification are strictly hierarchical: kingdom, phyla (in plants, divisions), class, order, family, genus, species. Linguists have not developed such a disciplined system, and even in biology the borderlines for all levels are somewhat subjective. There is no criterion to help decide whether a certain group of organisms qualifies, except for species, and the rule doesn't always work even for species. It is worth noting that a biological species shares certain features with language: both are groups of individuals able to communicate and to exchange information. Members of the same biological species can interbreed and thus exchange genetic information, just as individuals speaking the same language can exchange verbal information and communicate with others who use the same language.

As with biological organisms, in linguistics the most important taxonomic unit, the family or phylum, is represented by a group of languages that appear to share common origins. Until recently, this was the highest level of taxonomic unity generated by linguists. Increasingly, linguists are now recognizing groups of families that seem to share common origins, and the term "superfamily" has been coined to describe these. So far, however, there is not yet a classification of all languages that is unquestionably meaningful in evolutionary terms—a classification, that is, that takes into account both origins and history. A vague outline of its possible form is, however, just starting to emerge.

Linguistic families vary greatly in size and homogeneity; some include only a few languages, others a thousand or more, meaning that the degree of complexity also differs. At a critical point in a study with Menozzi and Piazza, I was lucky enough to see the proofs of a book, *A Guide to the World's Languages*, by Merritt Ruhlen, one of Greenberg's students. This work is the most modern, coherent, and systematic exposition to date of the history and philosophy of linguistic systematics and it includes a complete classification of five thousand languages. Access to it meant that we could organize biological data on the earth's populations in a simple hierarchy, and we treated it simply as a formal scale. Ultimately, we realized that the linguistic hierarchy actually had more profound significance than we thought originally, a subject we will return to a little later.

In Defense of Greenberg

Classifier's quarrels tend to be long, drawn-out affairs, and not only in linguistics; in the end, some kind of balance is generally reached that satisfies a sufficiently large number of experts, but sometimes only sheer exhaustion and boredom at the prospect of continuing largely pointless diatribes finally put a lid on the issue. The strongest disagreement currently surrounds the Amerind, partly because Greenberg's proposal is recent (1987).

I support Greenberg's views for various reasons. For one, I believe his synthetic approach is better than the binary one (which makes only two-by-two comparisons and limits itself to establishing the existence of a relationship, without analyzing the degree of affinity). Supporters of the binary system refuse, without comprehensible justification, to consider the similarities among the Amerind languages that distinguish them from all the rest of the world's languages—for example, the existence of only one system of personal pronouns. In Eurasiatic languages, the most common pronouns for the first and second person singular are *mi* and *ti*, in the Amerind languages, we find *n-* and *mi*. Personal pronouns are some of the most stable words and are therefore very useful in reconstructing affinities, particularly distant ones.

Some nonlinguistic studies also lend support to the notion that there were only three distinct migrations from Siberia to America, which correspond exactly to Greenberg's three families. One study is based on the shape of teeth, which can also be observed in ancient peoples and fossils. Another is based on genetic likeness.

Finally, there is a historical reason to trust Greenberg's conclusions. Greenberg has made an enormous contribution to many different areas of linguistics. He went into systematics more than thirty years ago, analyzing African languages, where there first reigned the same sort of confusion as today for America. Applying his method, Greenberg demonstrated that there are, in fact, only four families in Africa. This created an uproar among the linguists of the time, who were not inclined to abandon their previous positions. Today the whole debate is forgotten, and Greenberg's classification is accepted virtually by all. Those alive thirty years from now will be able to see whether history repeats itself.

A Brief Look at the World's Languages

If we also accept Greenberg's conclusions on Oceania, where the situation is much the same but less controversial, the world's five thousand languages can be grouped into seventeen families of varying size. Four are found in Africa, one in Australia, one in New Guinea, three in America, two in Europe, and the other six in Asia, with a certain amount of overlap near continental borders. Their geographic distribution can be linked to the history of the expansions of humanity and tallies well with what we currently know about migrations and genetic diversification.

Figure 7.1 shows the most recent geographic distribution of languages, published by Merritt Ruhlen a few years back.

Europe and Asia are not really separate units; it is better to consider the whole continent one large body, called Eurasia. The Ural mountain range, which is the traditional break-off point, does not extend as far as southern Russia. You could walk thousands and thousands of miles across the steppe lands from Romania to Manchuria on the Pacific coast in a landscape of tall grass that stretches virtually unbroken to the horizon.

One linguistic family, *Indo-European*, is distributed across Europe and southern Asia, with a break-off point around Turkey, where an *Altaic* language is used. The Altaic family stretches from most of Siberia and Mongolia right across to the Pacific Ocean, and several linguists also add to it both Korean and Japanese. It has spread quite recently, mainly by force of arms: the Turks arrived in Asia Minor at the end of the eleventh century and again in the fifteenth century, taking Constantinople in 1453, at which point Turkish started to supplant the local language, which was Greek. Another family, *Uralic*, straddles the Urals and is used

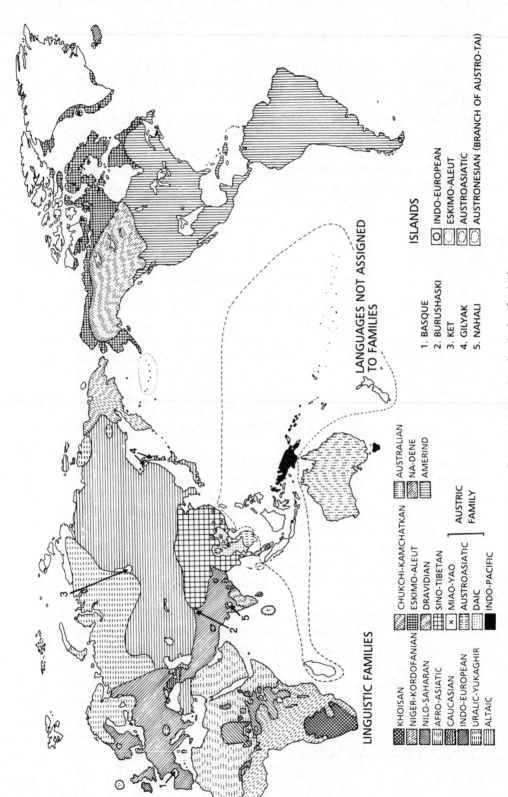

LINGUISTIC FAMILIES

KHOISAN	CHUKCHI-KAMCHATKAN
NIGER-KORDOFANIAN	ESKIMO-ALEUT
NILO-SAHARAN	DRAVIDIAN
AFRO-ASIATIC	SINO-TIBETAN
CAUCASIAN	MIAO-YAO
INDO-EUROPEAN	AUSTROASIATIC
URALIC-YUKAGHIR	DAIC
ALTAIC	INDO-PACIFIC

AUSTRIC FAMILY

AUSTRALIAN
NA-DENE
AMERIND

LANGUAGES NOT ASSIGNED TO FAMILIES

1. BASQUE
2. BURUSHASKI
3. KET
4. GILYAK
5. NAHALI

ISLANDS

INDO-EUROPEAN
ESKIMO-ALEUT
AUSTROASIATIC
AUSTRONESIAN (BRANCH OF AUSTRO-TAI)

7.1 Geographic distribution of linguistic families according to Ruhlen (largely based on Greenberg's classification).

in both Asia and Europe up around the Arctic Ocean, in a very cold region. There are actually two Caucasian families, rather than only one as shown in the drawing, and these are spoken in the mountains of the Caucasus close to the southern border of Europe and Asia.

In Asia, the *Sino-Tibetan* family covers the whole of China and Tibet. South of this region there are others, covering much more limited areas. The *Dravidian* family also covers a limited region. This family now comprises mainly the languages spoken in southern India, but there is reason to think that it was once spoken from Iran to Pakistan and throughout India.

Historical documents show that east of the city of Basra, which came to fame during the 1980s Iran-Iraqi war conflict, *Elamitic*, a language closely related to the Dravidian ones, was spoken from about forty-five hundred years ago until its disappearance some two-thousand years ago. The *Elamites* are mentioned in the Bible. Documents using their cuneiform alphabet made it possible to classify their language as Dravidian, a sensational discovery.

It is likely that the Dravidian languages were spoken from the western borders of Iran into and throughout the whole of India, and were first introduced by Neolithic cultivators from nine thousand years ago. It may be that the peoples of the Indus Valley civilization in today's Pakistan used Dravidian languages, but unfortunately we don't have enough written material to know for sure. Also known as Harappa or Moenjo Daro, after its major cities that were recently discovered and excavated, this civilization vanished some thirty-five hundred years ago for reasons that remain unclear; the date would appear to coincide with the invasion of Aryans, pastoral nomads using Indo-European languages who traveled south of the Urals through Turkestan and Iran. They probably brought languages such as Sanskrit to Pakistan and India. Nearly all the Dravidian languages spoken in India have disappeared. Only a few tribes in Pakistan, in northern India, and in much of the southern part of the Indian peninsula, continue to use them.

The arrival of conquerors who occupy large territories is often followed by the disappearance of local languages in all but fringe or particularly isolated areas, which are either inaccessible or of little economic interest, such as mountainous regions or certain islands. Many examples can be found of this principle at work in the geographic distribution of human languages.

One example is provided by the *Celtic* branch of the Indo-European family. These languages were spoken twenty-five hundred to three thousand years ago in central Europe and spread throughout the continent during the period 500 to 200 B.C. After this, the Roman era began. Southern and western Europe were taken over and Latin began to replace Celtic languages in France (one of which was spoken by the Gauls), in Spain, in northern Italy, and, a few centuries later, in Britain. Germanic languages spread to central Europe, and for a while the Celtic branch vanished from continental Europe.

Four Celtic languages are still spoken in areas far from its origins: in Scotland and Wales, where the Roman conquest was never completed, and in Ireland, which never came under Roman rule. Until recently, a Celtic language was also spoken in Cornwall, in the far southwest of England. In northwest France, the Bretons still speak a Celtic language, but this is a comeback: a number of Britons fled to Brittany during the Anglo-Saxon conquests after the fall of the Western Roman Empire, in the fifth and sixth century A.D.

In North Africa, the *Afro-Asiatic* family of languages is used, which also covers the Middle East, Arabia, and Ethiopia. It was once known as

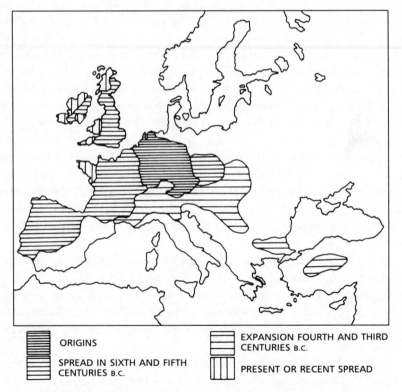

ORIGINS

SPREAD IN SIXTH AND FIFTH CENTURIES B.C.

EXPANSION FOURTH AND THIRD CENTURIES B.C.

PRESENT OR RECENT SPREAD

7.2 Distribution of ancient Celtic languages and of their more recent derivatives.

the Hamito-Semitic family; the Semitic branch includes Hebrew, Arabic, Aramaic, Assyrian, and many extinct Mesopotamian languages, along with some Afro-Asiatic languages still spoken in Ethiopia (such as Tigré and Amharan). In sub-Saharan Africa, *Niger-Kordofanian* languages dominate. This family is a rather unusual mixture of a small nucleus in the Sudan (the Kordofan region) and a very large group of languages spoken throughout western, central, and southern Africa. Sandwiched between the Afro-Asiatic and the Niger-Kordofanian families, we find the *Nilo-Saharan* languages. Last, in the far south of Africa, the *Khoisan* family, with its unique range of clicks, is spoken.

We know that America and Australia were occupied by peoples from Asia, and in this sense are appendages of Asia. In America, we find the three families already mentioned, Eskimo-Aleut, Na-Dene, and Amerind. Today's Australian Aborigines speak a multitude of different languages belonging to the family known as *Australian*, which was once even larger: each tribe spoke a different language, and there were five to six hundred tribes (many more than have survived into the present). The natives of New Guinea use languages from another family known as *Indo-Pacific*, which once must have taken in a vaster area. Both these families are probably very old, because Australia and New Guinea were first settled some sixty thousand years ago. It is not surprising that both comprise a great variety of languages that are not always easy to categorize given their likely long history of diversification.

In anthropological terms, the smaller Pacific islands contain three distinct human types: *Melanesians*, as dark as Africans, and with their thick, often curly hair, small body, and aquiline nose, they resemble the natives of New Guinea and inhabit neighboring islands; *Micronesians*, who occupy a relatively small group of more distant islands to the north of New Guinea and are a little different; and *Polynesians*, who are less dark and tend to be heavier set. Gauguin's brush and a series of Hollywood movies have given them the reputation for being contented. The Polynesians speak languages from a subfamily, called *Austronesian*, which, not surprisingly given the myriad islands of varying size in Polynesia, comprises 959 different languages. The Polynesians were excellent navigators, as demonstrated by the broad geographic spread of the subfamily, which is also found in both the Pacific and Indian Oceans. Austronesian languages are found as far west as Madagascar, near the African coast, which was occupied about one thousand years ago, as well as in the New Zealand islands, southeast Australia, the Hawaiian Islands, and on Easter Island, not far from the American coast. Among the Melanesians who inhabit islands such as Bougainville close to New

Guinea, there are languages from both the New Guinea and Austronesian families. The same sort of hodgepodge is found along the coastline of New Guinea itself.

This classification fails to find a niche for at least five languages, which do not appear attributable to any of the seventeen families. Basque is one, and the others are in Asia.

Families and Superfamilies: Eurasiatic and Nostratic

Many of the families mentioned are now accepted by all, or nearly, all, linguists, and similarities, albeit less marked, between groups of families are beginning to be recognized. These similarities are not always obvious. The problem is that languages change rapidly, and some linguists are convinced that there is no hope of establishing relationships from more than six thousand years ago. This conviction is supported by glottochronologists' opinion that after six thousand years the percentage of words in common has dropped to 10 percent; for a list of one or two hundred possible words, this leads to an excessive level of statistical error.

Glottochronological methods therefore cannot provide quantitative answers, particularly where long periods of time are involved. Focusing only on words with a high preservation rate and adopting other approaches, it has been possible to probe farther back in time, to the extent that a group of Russians and one American linguist have concluded that it is possible to establish the next level up in the classification by combining a number of families into a Eurasiatic "superfamily."

There are differences between the superfamily postulated by the Russian team and the American, who, once again, is Greenberg, the man who has made the most remarkable contribution of all time to linguistic systematics. Both group the Indo-European and Uralic families with the Altaic, but they differ over attribution of the others. Greenberg calls his superfamily "Eurasiatic," and it includes Japanese, Korean (considered distinct from the Altaic family), along with the Eskimo-Aleut and Chukchi families (in practice, the whole of north Eurasia, with ramifications in Iran, and the American Arctic regions). The Russians call their superfamily "Nostratic," and, in addition to most of Greenberg's families, it encompasses Dravidian, Afro-Asiatic, and part of the Caucasian family.

Greenberg used his own analytical method of simultaneously comparing numerous, most stable words, other parts of languages, and, in particular, grammar from many different languages. The following table gives a few examples of cognates; the hyphen indicates a suffix.

Language or Family	I	You	Plural Ending	Older Brother	Think
Indo-European	me*	tu,* te			med-*
Uralic	−m*	ti,* te	−t*	aka	mett
Mongol	mini	ti*	−t*	aqa	mede*
Korean	−ma				mit
Chuk-chi	−m	−t	−ti		mitelhen
Eskimo	−ma	−t	−t		misiyaa

*Reconstructed ancestral language form.

The meanings of these words are not always exactly the same as the column heading, but clearly derive from it. For example, *mitelhen* in Chukchi means "expert," *mit* in Korean is "to believe," and *mede* in Mongol means "to know." The extension of meaning across a wider semantic field is in itself an important feature of Greenberg's approach, since it allows us to probe farther back in time. Loans from other languages and coincidences may intervene, but Greenberg provides a number of good reasons why these sources of error do not seriously interfere with his conclusions. The list of words is limited to those least susceptible to change over time and therefore unlikely to be loaned from other languages. It is highly improbable that there should be multiple coincidences for numerous words drawn from a large number of languages. That is why Greenberg's mutlilateral method of simultaneously comparing numerous languages from the same family is so useful.

If we take languages closely related to those listed in the table, the similarities increase. For example, the first person singular pronoun in ancient Japanese is *mi*, older brother is also *aka* in Turkish, the Japanese of Ryukyu island, and in many of the Ainu languages. The Ainu once inhabited the whole Japanese archipelago, but today only live on Hokkaido and farther north in the Sakhalin islands.

The method used by the Russians is based on the reconstruction for each family of "proto-languages," or the hypothetical ancestral language from which the various modern languages descend (proto-Indo-European, and proto-Uralic, and so on). Obviously, only families for which linguists have already reconstructed the proto-language can be used, and this weakens the power of the method. Words cannot be reconstructed with precision, because it is often impossible to choose confidently among the many alternatives. The founder and most important exponent of the Russian Nostratic superfamily, V. M. Illich-Svitych, has also tried reconstructing the Nostratic proto-language from the proto-languages of its component families. He has even composed a poem in this hypothetical language, which perhaps would have been

NOSTRATIC SUPERFAMILY

EURASIATIC SUPERFAMILY

7.3 Greenberg's Eurasiatic superfamily and the Russian proposal for Nostratic.

spoken more than ten thousand or even twenty thousand years ago. Another Russian linguist, V. Shevoroshkin, believes that the Amerind languages should also be included within Nostratic.

Greenberg's conclusions and the Russians' are not really so disparate; bearing in mind the methodological differences and independent nature of both studies, the similarity between the conclusions drawn is astonishing. Greenberg believes that the Afro-Asiatic and Dravidian families diversified from those in his Eurasiatic superfamily earlier on, and that their degree of affinity is lower. The present differences between Nostratic and Eurasiatic probably will be resolved by a more general classification.

Other linguists have given these proposals a cool, and at times downright hostile, reception. Of course, very few academics are truly interested in the subject and have the knowledge required to deal with it. The great majority of linguists work on different, far more specialized issues, relating to much more recent periods of time, for which there is usually an ample documentary basis. Thus they tend to mistrust conclusions achieved on the basis of systematics, where both epoch and approach are unconventional. One can almost see the specter of an ancient taboo established by the French Society of Linguistics in Paris in 1866, which formally banned the study of linguistic evolution.

Nevertheless, the methods used would benefit considerably from a more quantitative approach, since this would render them more objective.

Was There Ever a Single Ancestral Language?

Other superfamilies have been suggested. Similarities have been recognized between the Na-Dene languages of North America, Sino-Tibetan, and a group of Caucasian languages, which happen to be those identified as resembling Basque. The geographic distance dividing these families (from Spain to North America) is enormous.

This superfamily, called Dene-Caucasian, was probably used throughout Eurasia before either Nostratic, or Eurasiatic, spread. The most compact families in geographic terms are those used by peoples that have spread more recently. There are several examples of ancient families or subfamilies being splintered and partially overlain by other groups expanding into the older language's territory. In this case, the Dene-Caucasian superfamily should be the oldest of all and date back more than thirty thousand years, if, as indeed seems very likely, Basque descends from the language spoken by the first modern humans to come to Europe, the Cro-Magnon. If the development and expansion of the Nostratic/Eurasiatic superfamily began some twenty thousand years ago, it is not surprising that Dene-Caucasian is forty thousand years old.

Other superfamilies have been postulated, and linguists are feeling their way toward the creation of a single evolutionary tree of language. But its full validation seems to be strictly for the future.

Now that a comprehensive linguistic classification able to group virtually all existing languages into a few initial branches is becoming feasible, one question springs naturally to mind: Has there ever been a period in the history of humankind when there was only one language? Many experts refuse even to contemplate the notion, claiming that linguistic evolution is too rapid to ever permit an answer. Joseph Greenberg (again!) attempted a reply, demonstrating that there exists at least one

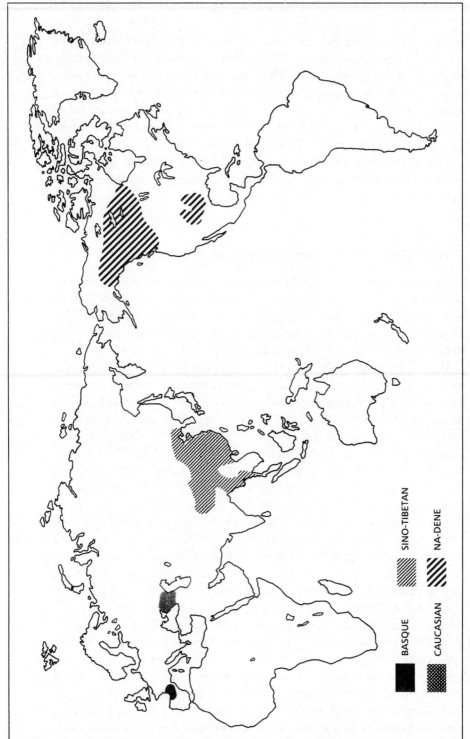

7.4 Modern distribution of the Na-Dene/Sino-Tibetan/Caucasian superfamily. This is probably the remains of a large, very ancient family, broken up by the subsequent expansions of the Nostratic/Eurasiatic superfamily.

word that seems to be common to all languages, the root word *tik*. Here are some variations:

Family or Language	Forms	Meaning
Nilo-Saharan	tok-tek-dik	one
Caucasian (south)	titi, tito	finger, single
Uralic	ik-odik-itik	one
Indo-European	dik-deik	to indicate/point
Japanese	te	hand
Eskimo	tik	index finger
Sino-Tibetan	tik	one
Austroasiatic	ti	hand, arm
Indo-Pacific	tong-tang-ten	finger, hand, arm
Na-Dene	tek-tiki-tak	one
Amerind	tik	finger

The only families for which examples of similarity have not been found are the main African ones, Khoisan and Niger-Kordofanian, which are probably the oldest linguistic groups of humankind. The semantic and phonetic variations are interesting. The latter can be quite extreme, an example being *tong* in the Indo-Pacific group. The meanings all center on the concept of the synonymy between *finger* and *one*, a fairly reasonable relationship because humans generally use the fore or index finger to indicate the number one. In addition to the fluctuation between *one* and *finger*, we see that *finger* sometimes becomes *hand*, or even *arm*, or the verb *to point*, while *one* becomes *single*. In Eskimo, *tik* describes the forefinger, whereas in the Aleutian islands (which lie off the coast of Alaska and are inhabited by fishing peoples who look like the Eskimos and share many of their customs, as well as their language), *tik* has come to mean the middle finger. The Indo-European verb *to indicate* or *show* (with forms *dik-deik* and *deik-numi* in ancient Greek, the root being the first part, *deik*) is probably the origin of the Latin *digitus* which gives us *dito* (meaning "finger" in modern Italian) and *digit* in English. There are always a few semantic gambols, accompanied by phonetic variation.

We have already said that *one*, *two*, and *three* are all strong survivor-words and are therefore especially useful when looking for links, even between distant languages. *One* comes first (not unnaturally). According to Merritt Ruhlen, a very common root is that corresponding to words for milk—all the best-preserved words relate to important issues in human experience. The almost universal root is similar to the English *milk*, in Greek and Latin, along with their derivatives, however, it has been replaced by another root, whose origin is unknown, *glac* (*galaktos* in Greek and *lac* in Latin). Ruhlen and another linguist, Bengston, have compiled about thirty interesting roots. These always relate to parts of the body (for example, knee, vagina, and eye) or important features of everyday life (such as water, lice and so on). Luckily, we have learned to control the lice problem, but there was a time when royalty—and other heads of state—had a hard time keeping them at bay.

For those intrigued by research into ancestral language, we provide some examples of universal roots (or etyma, as they are known), from Bengston and Ruhlen's work:

Family	Who	Two	Arm	Vagina	Water
Khoisan	!ku		//kan	k″a	
Nilo-Saharan	kukne	ball-	-kani	buti	kwe
Niger-Kordofanian	ki*	bala	kono	butu	
Afro-Asiatic	k(w)*	bwr*	-gan*	put*	ak′w*
Nostratic/Eurasiatic	ki*	pala*	kon*	poto*	ak a*
Dene-Caucasian	ki*		kan*	puti*	ok a*
Austric	o-ko-e	mbar*	xeen*	betic*	
Indo-Pacific		boula	akan		okho
Australian	kuwa	bula*		puda	gugu*
Amerind	kune	pal*	kano	butie	akwa*

*Reconstructed ancestral language (proto-language) sound.

It is worth remembering that an Italian linguist, an exceptional polyglot and renowned academic, Alfredo Trombetti, published work at the beginning of the twentieth century suggesting that all languages share the same origins, and was mocked by his fellow scholars. Now that this idea is gaining ground, an English translation of his work is planned.

It will be a long time before any kind of universal, or even broad, agreement is reached on such a thorny subject. And two question marks will remain: If there was a single language, when did it exist? A half-reply is: Obviously, before modern humans' first diaspora, which means at least sixty thousand years ago. And again: When did humans first start speaking?

When Did Language First Appear?

That language appeared overnight, as it were, and immediately became as sophisticated as it is today, would be hard to believe. There is, however a small piece of evidence that in the oldest species of humanity, *Homo habilis*, the biological basis for some primitive language form already existed.

We know there are areas important for language in the brain, somewhere behind the eye, because when these are damaged by injury or stroke, the ability to produce and comprehend language, and to write, is impaired. These areas (known as Broca and Wernicke) are found in the temporal region of the brain's left hemisphere and make the cranium slightly asymmetrical, with the left side being slightly larger. This asymmetrical form is already found in our most intact *Homo habilis* skulls from more than two million years ago, but is absent from the apes closest to humans.

It has now been shown that chimpanzees and gorillas can learn the meaning of hundreds of words and use them in quite long, if grammatically simple, sentences. These distant cousins of the human race are able to grasp the meanings of symbols, but are incapable of producing human sounds. We have to use special systems to communicate with them, such as sign language and computer symbols.

Perhaps in the earliest members of the species *Homo*, voice had already begun to develop. But a huge amount of time was needed to achieve the breadth and complexity of sounds reproducible and comprehensible today, and for special brain areas, dedicated to memorizing our enormous vocabulary and generating and comprehending our complicated linguistic structures, to develop.

The braincase of these ancestors was still very small (only about half to a third the size of ours) , and there is no doubt that the large subsequent growth must have served to some extent—and perhaps above all—to contain the structures for language. The human brain ceased to grow some three hundred thousand years ago, but it must have taken a long time after that for humans to reach their present level of articulation.

It is thought that the Neandertals, who lived between thirty-five thousand and three hundred thousand years ago, may have been unable to speak as well as we do; because their larynx and pharynx did not develop enough. These are soft organs that leave no trace after decomposition, so the notion is based on information attained indirectly by studying parts that have remained. Whether the idea is right or not, this seems a cogent proposition.

There are other reasons to believe that acquisition of a high level of language with broad vocabulary and complex syntax is a recent phenomenon. All languages share one common feature: they are very similar in terms of degree of complexity. The languages of peoples generally considered primitive are even richer and more complex than ours (note that over the years English has almost stopped conjugating verbs and nearly all the Latin noun declensions have been lost during its transformation into Romance languages). The biology of those speaking these languages is almost identical; there is no difference whatsoever in the ability of people to learn any language, be they English, Australian aboriginal, or Amerind. As mentioned earlier, any normal person can learn any language if he or she starts young enough, because the ability to master a language perfectly is lost very early.

Modern Humanity's Most Valuable Tool

Complex language also means considerable intelligence. We know that during the period from sixty to one hundred thousand years ago there were important improvements in stone tooling, and the spread of humankind across the globe began (terminating sometime between ten and thirty thousand years ago). Once the four corners of the earth had been reached, *Homo sapiens sapiens* continued to increase in number, but could no longer expand quite so easily. Food production became a necessity as the hunter-gatherer lifestyle no longer guaranteed nourishment for all.

A fast, precise method of communication must have been of enormous help in the spread of humanity. It was certainly crucial to sending scouting parties ahead to report on the best routes and climate conditions, but also to adapt to new climates with different flora and fauna, and perhaps full of difficulties and unknown dangers. The ability and inventiveness that allowed the use of newly available materials to build new arms, plan explorations and migrations, build dwellings, and make clothes suitable for various climates, were undoubtedly sharpened by the ability to communicate.

Culture that is both complex and volatile tends to differentiate locally. It is not surprising, therefore, that archaeologists have noticed another diversification, parallel to the linguistic one, that was especially pronounced in the last fifty thousand years: the considerable differentiation of tooling industries and the adoption of new materials, such as bone, ivory, and wood. That this has been taken as an indirect proof of the existence of more refined languages that begin to alter from place to place seems reasonable as well.

Biological and Linguistic Evolution

Since their earliest beginnings, the disciplines of linguistics and biology have exchanged ideas, albeit on an informal level.

Around the midpoint of the last century, Charles Darwin explained biological evolution as the consequence of the natural process that generates living creatures by trial and error. The most successful organisms are those that prove most functional because they are suited to the environment and are automatically selected because they multiply more than the others.

Nature's "try and try again" method consists of constantly proposing new spontaneous mutations (a concept unclear in Darwin's days). Natural selection is the gauge that discards harmful mutations and promotes beneficial ones, simply because the latter multiply and spread more than the former. On the basis of these key concepts, Darwin explains the multiplication, transformation, and diversification of living organisms and illustrates them in *On The Origin of Species* with hypothetical examples of evolutionary diagrams of species. Not long after, a linguist, August Schleicher, published the evolutionary diagram of Indo-European languages shown in Figure 7.5.

The modern view is a little different, and the tree's value is now mainly historical. A recent survey involving two famous linguists, Isodore Dyen and Paul Black, and the statistician Joseph B. Kruskal looks at cognates through a classic list of two hundred words in eighty-four Indo-European languages, and differs somewhat from other accepted views, including that presented in Schleicher's tree. Dyen, Black, and Kruskal's work

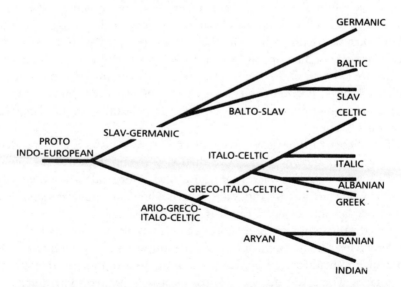

7.5 One of the very first evolutionary trees in the history of linguistics and biology. Published in 1863, it shows the renowned linguist August Schleicher's view of the origins of Indo-European languages. Subsequent diagrams are not necessarily any better, but there are difficulties in agreeing with all of Schleicher's earlier branchings

recognizes the existence of groups of Celtic, Italic, Germanic, Balto-Slavic, Indian, Greek, Iranian, Armenian, and Albanian languages (note that Schleicher's diagram omits Armenian languages completely): it finds rather weak links between Italic, Germanic, and Balto-Slavic languages; but (in contrast to Schleicher) finds no similarity at all between Italic and Celtic languages, nor between Indian and Iranian ones or any of the above-mentioned nine groups. The nonidentification of an Indo-Iranian group by Dyen, et al. is the major departure from the conclusions accepted by the majority of traditional linguists.

Theoretical models of evolutionary diagrams don't always help in representing the diversification of languages, because neighboring peoples may swap words and influence each other reciprocally in a form of "migration" not comparable to that in biology. Evolutionary studies of species don't have to take migratory exchanges into account because by definition the *interbreeding between individuals of different species cannot generate fertile offspring*. When studying the evolution of populations of the same species, which can have migratory exchanges, the validity of a tree depends on the exchanges being small. There are other important differences between migratory exchanges in language and biology.

Words from other linguistic groups (known as loan words) are constantly entering foreign languages; this is a source of irritation to purists who wish to avoid contamination. However, they are a measure of the inevitable cultural influence one people has on another. The phenomenon occurs so frequently that one theory of linguistic evolution states that words migrate from their origin outward with a concentric wave movement, like ripples from a stone thrown into a pond. Linguistic maps are used to show the areas reached by a word or expression or other linguistic phenomenon that spreads this way. The name for these wave lines on a linguistic map is isoglosses, meaning "equal language contours."

This theory of the spread of words and expressions as an independent phenomenon, known as the wave theory, is the antithesis of the evolutionary tree, in which each language evolves independently. Obviously, both models are useful and valid, but in different circumstances. One model does consider both, but it is very complicated.

Exactly the same phenomenon occurs in biology. The evolution of different species and populations of the same species that rarely or never exchange individuals can be described using an evolutionary tree, as can languages where the number of loan words is limited. There are also theories in biology of "isolation by distance," which predict (correctly) that geographically close peoples exchange individuals with greater fre-

quando

quanno

7.6 Isoglosses and geographic distribution of two linguistic expressions in Italy. In the northern part of Italy, the personal pronoun habitually procedes the noun (as in the English "my son," which is the meaning of the phrase), whereas in the south the personal pronoun, *mio*, comes last. *Quando* and *quanno* are alternative pronunciations of the word for "when."

quency and therefore become more alike genetically than peoples separated by larger distances. The theory that genetic distance between two peoples increases in relation to geographic distance has proved valid for languages as well.

In linguistics, there are phenomena comparable to mutations: changes in vowels and consonants, abbreviation or extension or words. (In some languages, words often double. This is particularly noticeable in the Polynesian languages and various parts of Africa, where, for example, *pili-pili* is used for "chili"). Semantic variations are commonplace, as are phonetic modifications, but grammatical changes are rarer. Without mutations, language would not change. Linguistic innovations generated in this way do not have to pass the test of natural selection. They are, however, subject to a form of selection we can define as cultural. This selection applies to all cultural phenomena, of which language is undoubtedly one of the most important forms. It is, therefore, reasonable to speak of *cultural selection*.

Natural Selection, Cultural Selection

It is nature, in the sense of living conditions in their widest possible application, that practices natural selection. Innovations brought about by mutation are checked, tried, assessed, and either adopted or scrapped. Cultural selection is done by human communities; if we are faced with a new word, we size it up and then make our own decision.

Language on the whole has eminently practical applications and is aimed at promoting cooperation and information exchanges between human beings; this is proved by the multitude of words for instruments and specialized work tasks. Naturally, it is often technological innovation that stimulates the invention of new words, required to define objects that didn't previously exist. Some terms catch on, but others fall by the wayside: "horseless carriage" has been dropped in favor of "automobile." just as "airplane" has ousted "flying machine" (whereas "washing machine" has been accepted).

Some words undergo a full-fledged trial by natural selection, which is completely independent of cultural selection. Natural selection involves one set of individuals surviving better than another, or reproducing more. In certain exceptional circumstances, the ability to pronounce a word is a question of life or death. In the Sicilian Vespers (1282), a revolt broke out against the French occupying force in Palermo. The folk story is that, while searching a local girl, a French soldier tried to take advantage of the situation in a way that intensely irritated the girl's family, who were present. This episode provoked an uprising and expulsion of the French from Palermo. Modern historians explain the uprising in more clinical terms as just one episode in the Aragon/Angevin struggle for control of Sicily. To identify their enemies, the Sicilians used to make them say *ceci* (chick peas). Inability to get their tongues around the word cost many French people their lives. The Bible recounts a similar story: Some Jews were able to recognize and kill members of another hostile Jewish group by their inability to correctly pronounce *shibboleth*. Overall, one can see only advantages in being multilingual.

Language evolution contains much cultural selection and very little natural selection. In general, shorter words are more popular than long ones, and in certain areas the preference for shortness is very pronounced. This may be why the French have dropped the last syllable in so many words, and why their Catalan neighbors are heading the same way. (Like many Latin derivatives, including Italian, in French the emphasis was once on the penultimate syllable of words. When the last syllable was dropped, the penultimate became the last one. The habit of emphasizing the last syllable is now so engrained that the modern French often stress it even when they shouldn't. Modern Catalan, known as Català, is very similar to the ancient language spoken in the south of France, the Langue d'Oc and adopted a similar custom.)

Language and Brain

Certain evolutionary processes seem characteristic of language, and are not found in other areas of evolution, such as biology or culture, or at least not so obviously. One of these is lexical diffusion—the spread of an innovation from one word to other similar ones. For example, in English the past tense can be indicated in several ways, either by adding the suffix *ed* (changing *love* to *loved*) or a different suffix, or even by changing root. These last two types of verb are termed irregular, and examples are *find* (*found*) and *go* (*went*). The number of regular verbs has increased considerably since medieval times; this process of simplification has accompanied the various transformations of English down the centuries.

Cases of lexical diffusion where a given form or, more precisely, a particular model is progressively adopted by analogy are fairly frequent. In Italian, we find that the letter *n* before *s* followed by another consonant (known as the impure *s*) has been dropped. So *inspiration* has become *ispirazione*, *transport* has become *trasporto*, and *Institute* has become *Istituto*. French, and the Latin original, of course, have kept the *n*, so that the French *translation* is *traslazione* in Italian.

In a talk he gave as president of the American Language Society, the Philadelphia sociolinguist William Labov stated that the recognition of lexical diffusion by William Wang of Berkeley has constituted one of the greatest steps forward in linguistic evolution in recent times. Labov is now examining whether a phenomenon recently appearing in many languages, and with singular emphasis in English, can be considered an example of lexical diffusion. It is one of the most interesting developments in the English language since medieval times. Called "the great vowel shift," this complex phenomenon is a major culprit in the marked divergence between English spelling and pronunciation. When this shift began, there were a limited number of vowel sounds in English (about seven, in line with other Latin-derived languages, where the five written vowels have seven sounds because *e* and *o* can both be pronounced either open or closed). Currently, English has about twenty distinct vowel sounds and a large number of diphthongs.

In the Middle Ages, many words such as "bite," "mite" and "white" were pronounced like an Italian or Spaniard would pronounce them today. It is not easy to render this pronunciation with modern English spelling, but it would be something like that of words "beetay," "meetay," "wheetay," except that the final *y* would not be heard. But starting in the fourteenth century and going through a series of stages, more or less the same for all words, the last vowel was dropped and the first became the diphthong "ai," like in the modern "received pronunciation" (British upper class, and BBC of some years ago). In phonetic spelling, the original *i* became first *ii* then *ei* (as in wait) and finally *ai* (as in white). In some parts of the UK the various older pronunciations are preserved; in others, the pronunciation has gone one step further or in other directions. For instance, in the London dialect (Cockney) and in Australia (many of the first Australian settlers were from the London jails) the diphthong of "bite, mite, white" had become *oi*. Lexical diffusion assured, of course, that all similar words acquired the same sound. In these last regions, another vowel shift affected words like "mate" (meaning pal), "rate," "wait" and "bait". It seems as if, in the English upper classes, these words are still at the "ei" stage, while in Cockney and Australian they have reached the next stage, "ai".

A well-known linguist, E. Sapir, has called this unidirectional parallel variation "drift." Unfortunately the meaning is different from that in genetics. Genetic drift does not have a precise direction; it moves with equal probability in both possible directions (increase or decrease in the frequency of a gene form).

The linguistic use is more correct, because *to drift* means "to be drawn by a current," and currents have a fixed direction. That a linguist's definition of a word is more accurate than a geneticist's is not entirely surprising, because it is the linguist's job to be precise about language. To emphasize the gap between the genetic and more common usages of *drift*, Motoo Kimura, a geneticist whose contribution to the mathematical theory of genetic drift has been very significant, suggests we talk of "random genetic drift," which is clear, but rather long-winded for everyday use.

With a reasonably ample definition, it seems that lexical diffusion can be extended to include the classic "phonetic laws" established in the nineteenth century. "Phonetic laws" describe the regular way in which sounds change. They have been proved and applied on many occasions particularly in the nineteenth century. Among these are the famous Grimm laws (defined by the excellent linguist Jacob Grimm, who together with his brother also wrote the famous fairy tales). For example, *p*, *t*, and *k* in ancient languages (Greek, Latin, or Sanskrit, that is) become *f*, *th*, and *h* in English, and *f*, *d*, and *h* in High German. So *pater* is *father* in English, *fadar* in Gothic, and *Vater*, pronounced "fater" with a long *a* in modern High German). A group of late-nineteenth century linguists, the neogrammarians, were convinced that the rules for changes in sound are perfect, and that all exceptions can be explained. This is an exaggeration, but it must be admitted that sounds change with astonishing regularity. It is reasonable, therefore, to seek an explanation in some biological substrate.

Another well-known linguist, Noam Chomsky, recognized that language has a profound structure, manifested by humans' ability to perceive subtle differences between superficially similar-sounding phrases. He suggested that the human mind has an innate capacity for language, meaning that in his opinion there is a unique and special biological basis to language. It is easy to be less enthusiastic than his more ardent disciples about many details of Chomsky's theory, but there is undoubtedly truth in the affirmation that the human mind has a predisposition for language. Animals, even those closest to us, don't have this predisposition and probably will never achieve our degree of efficiency and complexity of usage. This unique aspect of the human mind has so far manifested itself in two ways: the enormous interest displayed by a normal child in learning language in order to be able to communicate with adults and other children, and the presence of special abilities with a genetic basis, which allows us to absorb the finer details of the use of language and its structures.

Children who have no contact with other humans from whom to learn language in early life later lose the ability to learn and are doomed to remain partially or totally speechless. A suitable social context is imperative to activate the innate learning mechanism. There are many examples of so-called wolf-children—people cut off from other humans for many years after birth, because they are brought up by animals (wolves, bears, sheep or pigs, for example) or for other reasons—who have partially or entirely lost their ability to learn to speak. The most recent case is that of an American girl, Genie, whose father kept her tied up in a locked room for several years, which prevented her from learning language. When this dreadful abuse was discovered and stopped, various experts studied Genie and tried to teach her to speak, with only partial success.

There must be a critical period in which children have a strong predisposition to acquire language and a powerful urge to do so, both clear signs of a biological basis, forming part of a normal human's psychological makeup. Another critical period, already mentioned, which for most people terminates at puberty, fosters the ability to become fully bilingual and, in particular, to perfectly reproduce sounds that differ from those of the native language. For example, French and Italians find it particularly difficult to learn the two *th* sounds (as in *the* and *Smith*) in English. With the exception of a few lucky individuals, the only way to perfection is to get used to pronouncing unfamiliar sounds well before puberty.

The production of unaccustomed sounds is not the only test every child sails through, while most adults find it an insurmountable hurdle. During life— and particularly when young—humans learn the multitude of rules governing all languages. Virtually everyone is able to learn these, although there are exceptions. Some people's use of language is severely handicapped by damage in certain areas of the brain or inherited genetic lesions. These handicaps have been studied only recently, and the number of fully analyzed cases is still limited. In addition to serious disorders involving hearing and/or speech impairment, or aphasia (the total lack of ability to understand or produce language), there are less severe ones, such as dyslexia and dysgraphia, respectively the inability to read and to write correctly. A number of disorders are highly specialized, such as the inability to conjugate verbs or nouns, or both, and these are naturally of great interest to specialists.

In a recent case in Canada, several members from different generations of a large family were unable to form the plural of unusual nouns correctly. If we were to ask them "How do you say more than one hat?" we would receive the reply "hats" without any hesitation, because the word is common. If, however, we were to explain that there is a certain animal called a "wombat," which they had never heard of, and ask them to supply the plural, we would not be able to make them answer "wombats." This means that the rule for forming plurals is missing, and it reveals the probable existence of a remarkable gene that controls the ability to form plurals or conjugate verbs. Molecular genetics may eventually allow us to identify the gene responsible for this specific defect.

The human brain usually recognizes these rules and applies them subconsciously, probably to reduce the effort required to communicate and make speech more efficient.

Is There a Relationship Between Biological and Linguistic Evolution?

We have seen how the evolutionary tree of languages is full of uncertainties and contains lacunae. The one for genetic origins is more reliable although it too runs the risk that future research will weaken or replace some of the links established to date.

However, both are at a point where we can start asking the question "Is there a parallel between linguistic evolution and genetic evolution?" In an article published in 1988, Alberto Piazza, Paolo Menozzi, Joanna Mountain and I provided an initial answer, based on our analysis of the data used to build the evolutionary tree of human populations.

We had listed the populations whose genetic data we studied on the basis of a linguistic classification, because that was the simplest and most complete means of ordering the large amounts of data at hand (piles of genetic data for fifteen hundred populations). The approximately five thousand languages in existence correspond closely to the nations and native tribes of today. The use of a linguistic classification of populations under study made it easy for us to check whether there was a relationship between the genetic and linguistic trees. And the answer was yes.

We reduced the fifteen hundred populations for which we had genetic data to forty-two by regrouping them along geographic and ethnic lines. We sometimes drew on linguistics to decide the makeup of a group, but we made sure that this could not invalidate our conclusions. It should be said that the linguistic criterion for grouping language often gives similar results to a geographic one or that to be deduced from physical and cultural similarities. For simplicity, the forty-two populations are here reduced to twenty-seven. For example, the majority of European populations had a very similar range of genetic differences compared with those distinguishing them from peoples of other continents. So we lumped the six European populations together as one group and left out only the Lapps, who were very different.

If we match the families and superfamilies listed by Ruhlen with our genetic tree, as in Figure 7.7, we see that on the whole the families unite branches of the tree that are close together, and have divided only recently. There are only a few exceptions to this rule, and these are all easily justified. It is also of note that the largest superfamilies, such as Nostratic and Eurasiatic, approximately coincide with a major branch. This is true, too, for the other superfamilies, such as Austric, which takes in the large Austronesian group and includes the Malayan-Polynesian languages.

Hoping to be able to fine-tune this genetic tree, we are watching future improvements of the linguistic tree carefully. However, we are already sure that the tie-in between the two is not coincidental. This view has been contested by some linguists in the opposing faction (Americanists in the anti-Greenberg coalition), who are not amused by

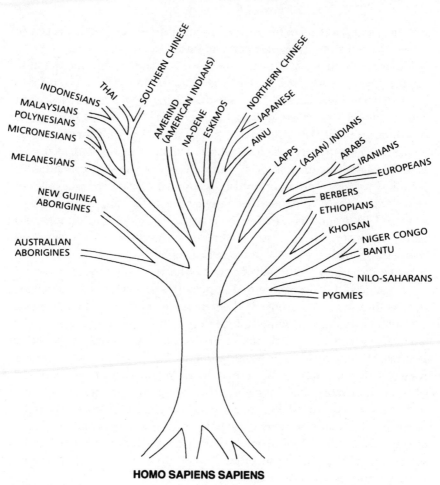

HOMO SAPIENS SAPIENS

7.7 The relationship between the genetic tree of the world's major populations and linguistic families.

the fact that the similarities among the Amerind in the genetic tree reflect the linguistic groups identified by Greenberg. We have, however, demonstrated that the chances of the similarity between the genetic tree and linguistic classification being due to chance are negligible. This has been tested today with two independent statistical methods.

The Exception "Tests" the Rule

Most people have heard the saying "The exception proves the rule." This is silly, because the exception, if proved correct, destroys the rule;

but almost every rule has its exceptions. The saying originally was "The exception *tests* the rule." Some exceptions are so fundamental that they destroy an entire theory, but others, if backed by a suitably convincing explanation, may reinforce it.

Let us look at the most important exceptions in our results. The Ethiopians comprise a number of different ethnic groups and have many more languages. They are one of the forty-two genetic groups emerging from the fifteen hundred populations studied, and are classified as African, genetically speaking, even if a closer look reveals that they are special Africans with a high level of genes of Caucasoid (white) origin. In fact, we can call them an admixture of African and west Asian (Arab) genes. The two groups contribute respectively about 60 percent and 40 percent of the genes. But linguistically speaking, they are closer to the Arabs, because they generally speak languages from a family (Afro-Asiatic) covering northern Africa, Arabia, and the Middle East.

The mixed genetic makeup and use of Afro-Asiatic languages reflect the history of the Ethiopians, who for a long time had close contacts with the Arabs. In and around the earliest Christian times, there was an empire that took in both regions. Its capital was first at Saba (Sheba) in Arabia and later at Axum, in Africa. According to Ethiopian tradition, Makeda, the Queen of Sheba, visited King Solomon and had by him a son, Menelek, founder of the Ethiopian dynasty, which has only recently been overthrown. The Bible tells of these events. Given their relationship of more than three thousand years, it is not surprising that the Africans and Ethiopians sit together on the genetic tree, but alongside the Arabs on the linguistic one.

Lapp history is in many ways similar. Genetic data position them among the Europeans (although they are extreme outliers in the European tree), whereas linguistic data include them in the Uralic family, which straddles northern Europe and western Siberia. Further genetic analysis shows that in fact they are genetically intermediate between Siberians and Europeans: The two groups have undoubtedly mixed to the extent that they are now over 50 percent European and often fair-skinned (even if dark Lapps are by no means rare). They have lived in Lapland, in northern Scandinavia, for at least two thousand years and probably much longer. They have adapted to their environment extraordinarily well and fish, hunt wild reindeer, and also raise it as livestock. Two thousand years ago, they had already started to use wooden skis to cross the frozen wastelands in winter.

A different example where there is no clear genetic mix is provided by the Tibetans (not shown in Figure 7.7), who come from northern China and speak Sino-Tibetan languages. In our tree, these languages are associated with the southern Chinese, who are generally very distinct from the northern Chinese. Unfortunately, we couldn't include the northern Chinese in the original genetic tree, because we didn't have enough reliable data. We do know, however, that the Chinese language belongs to the Sino-Tibetan family. It was spoken in northern China more than three thousand years ago, and spread southward from the unification of China onward, around twenty-two hundred years ago. Rather than reflecting their origins, therefore, the Tibetan's linguistic difference from the southern Chinese reflects subsequent, political events.

The Tongues of the Conquerors

Looking carefully, we find that these are not really exceptions at all, particularly if we bear in mind that in linguistic evolution there is a special mechanism by which one language can completely replace another in certain sociopolitical circumstances. This is entirely different from the loan mechanism that allows individual words to migrate from one language to another.

A good example is the recent history of the languages of America. Until the time of Columbus, Amerind languages were spoken in much of North America and throughout Central and South America. European colonizers brought four languages with them: English and French in most of North America, and Spanish in Central and South America, except for the eastern part of South America where Portuguese is spoken. In addition to conserving their own languages, the colonizers passed them to the natives. Fortunately, many of the original languages have survived and can be studied, but a number are now disappearing fast, and in two or three hundred years' time few of the six hundred languages surviving today will still exist.

r replacement to occur, the newcomers generally have to be numerou but, above all, powerful. The invention of modern weaponry and occupying armies has furthered language replacement over the past few centuries, but the phenomenon is not limited to recent times. We have seen how the arrival of conquerors speaking Indo-European languages in Iran and India was instrumental in the replacement of Dravidian languages. Although we lack hard-and-fast historical proof, it looks as if the same happened with the Celtic conquerors of Europe. The languages they imposed gave way to Latin in western Europe, and also in England where Latin was later replaced by the languages of the Anglo-Saxon conquerors.

Conquerors have not always imposed their languages. The Romans allowed Greek to remain the official and spoken language in Greece, Turkey, and southern Italy, because it was considered a prestigious language. The Franks, a Germanic tribe that invaded northern France at the end of the sixth century A.D. after the fall of the Roman Empire, did not suppress Latin, which had already begun to evolve into modern French. Another Germanic tribe that perhaps came from Sweden, the Longobards or Lombards, conquered the better part of Italy between the sixth and eighth centuries A.D. and imposed power and law, but not language. Modern Italian still has a few Longobard words, linked to the customs and laws of the time: Some place names (such as *Sala* and *Farra*), first names, and a greater number of surnames (*Adimari*, *Anselmi*, *Alberti*, *Berlinguer*, and so on).

Language replacement is an extreme example of rapid linguistic evolution. One language can replace another in just a few generations' time. In Africa I once came across a pygmy village where the list of questions on demography I successfully used in other villages were understood only by the older members of the village. This pygmy village was unusual in other ways, too: It was on the edge of the area inhabited by the pygmies of the Central African Republic and built in the middle of a field, instead of in or near the forest. What took me most by surprise, however, was the sight of hens pecking around the huts; they were the first domestic animals I had ever seen among the pygmies.

Darwin's Prophecy

At a certain point I realized with a mixture of emotion, pleasure, and some embarrassment that our observations on the strong similarity between the genetic tree and the linguistic one had been presaged by Charles Darwin. Here is a quotation from *The Origin of Species* (1859): "If we possessed a perfect pedigree of mankind, a genealogical arrangement of the races of man would afford the best classification of the various languages now spoken throughout the world, and if all extinct languages and all intermediate and slowly changing dialects, were to be included, such an arrangement would be the only possible one." I was reminded of this prophecy by Franco Scudo, an historian of evolution; I was embarrassed because it had completely slipped my mind.

Why should there be a similarity between linguistic evolution and genetic evolution? The explanation is simple. During modern humanity's expansion, breakaway groups settled in new locations and occupied new continents; from these, other groups broke away and traveled to more distant regions. These schisms and shifts took humanity to very remote areas where contact with the original areas and peoples became difficult or impossible. The isolation of numerous groups had two inevitable consequences: the formation of genetic differences and the formation of linguistic differences. Both take their own path and have their own rules, but the sequence of divisions that caused diversification is common to both. Their history, whether reconstructed using language or genes, is that of their migrations and fissions and is therefore inevitably the same.

It could be objected that the history of the world is not made solely of fissions of peoples. There have been clear-cut and significant breaks, such as the occupation of new continents and islands, where geographic discontinuity encouraged genetic and cultural discontinuity. But even where geographic continuity favors exchanges, there is still a correlation between genes and language, although the exceptions are naturally greater. The factors governing the formation of differences between groups—isolation and migration, that is—act on DNA and language in similar, although not identical ways, even in the absence of clear-cut geographic discontinuity.

EIGHT

Cultural Legacies, Genetic
Legacies

Contact with the pygmies came as a shock to me. The difference between my way of life and the one that comes so naturally to them is incredible. They live in the roughest of huts built with their own hands in a couple of hours, in the gloom of the tropical forest where the sun's rays rarely penetrate the high vault of dense foliage. They normally live in groups of no more than twenty or thirty people. They hunt their meat daily, using rudimentary weapons and ancient stratagems to surprise their prey, and patiently gather a huge variety of leaves, roots, and fruits for which western botanists don't always have a name. And yet it takes only a few days in their company to realize that we are the same people and, furthermore, that they are profoundly likable. It is true that there are physical differences between us, but they are polite and dignified, answer a smile with a smile, and can quickly tell a friend from an enemy. Many of them are very brave, some have made important discoveries regarding animal behavior, and other have invented new medicines and hunting techniques. They have survived infinite dangers, trials, and hardships.

It takes great courage to stand between the legs of a live elephant and kill it with a wooden spear. All the African elephant ivory of the past few centuries was hunted by pygmies—the farmers merely carried the tusks to the merchants' ships. It is the pygmies who catch the truly rare animals of the tropical forest seen in zoological gardens. From an ecologist's standpoint, this may not be to their credit, but as long as the forest was in the hands of the pygmies no rare animal ever risked extinction. It is the rifle, brought by white settlers, that has decimated the fauna of Africa. Even today the pygmies rarely use rifles, and when they do, the weapon is always borrowed. It took me a little while to realize that there was no difference in emotion or intelligence between ourselves and these small people, somewhat bewildered by the large, pale-skinned individuals buzzing around with mounds of instruments and odd devices. Everyone on our team, which spent hours interviewing and studying these beings from the inner forest, quickly developed the same liking and admiration. It was a lesson to me to see the respect with which a great anthropologist such as Colin Turnbull, with whom I visited many camps, treated them. On the very first day, I noticed how he bent his long-limbed six-foot body to speak with a tiny, elderly pygmy woman in a cultivators' settlement, a place where the pygmies are often treated with arrogance and contempt, like stupid servants. His Scottish ancestors would have addressed a queen no differently.

It is natural to wonder why this huge difference between our way of life and theirs has come about. The biological differences are obviously striking, and equally obviously superficial. The pygmy economy is totally different from ours, but this alone cannot account for the degree of diversity. The explanation must lie in a radically different cultural legacy dating back thousands of years, which is well suited to the conditions of life and very resistant to change. Then again, why should they change? The only reason is that we are stealing their forest from around them, because it suits us, and we don't care in the slightest that we are destroying their way of life without being able to offer them an alternative one; and if we could, it would be infinitely worse than what they are losing.

Genes, Appearance, and Behavior

During the nineteenth century, "national character" was a favorite topic for debate. What causes it? Is it genetic or is it cultural? As a student, I lived among people of different origins. It was easy to guess the nationality of my fellow students; this was naturally based on physical observation and scarcely considered behavior. I began to wonder whether Europe really does contain all these different races, after all. I decided to pinpoint the factors I relied on to identify nationality. Hair and eye color were of immediate help, of course, but other features were much more important: shoe shape, the style and color of clothes, and, above all, haircut, all of which are entirely cultural.

If we compare Rh, ABO, and other known gene frequencies among European peoples, we find that the differences are minimal. The Germans and French have fought and hated one another over the centuries, but in most ways tend to be very similar. There are greater differences within France between those from the southeast or southwest and those from the northwest, for example. On the whole, Europe is exceptional in that it is extraordinarily uniform in genetic terms, at least in comparison with other continents.

A major question mark, however, remains. Hair, eye, and skin color clearly vary considerably between north and south. This is biological and hereditary, not cultural. So why are these visible biological differences greater than those observed in blood groups or other genes? And if there are genes controlling hair color, why shouldn't there be genes controlling self-discipline or sense of humor? We don't know the answer, and modern genetics is certainly not in a position to analyze such elusive features. Perhaps things will be different in twenty or thirty years' time.

We have already explained differences in skin color as the outcome of environmental and climatic adaptation. Hair and eye color may simply follow suit with skin pigmentation, except that association is not hard and fast—there are people with black hair and blue eyes (a rare, attractive combination). These features are important in the choice of a marriage partner and have a selective advantage, because fair coloring seems to be most sought-after where it is least frequent, and vice versa. This is sexual selection, which makes us desire less frequent combinations. Exceptional pigmentation types such as albinos (about one of every ten thousand births) have often enjoyed great success, and have even been thought to have divine origins.

We cannot exclude the possibility that there is a genetic component in the behavioral characteristics that incline people to Nazism or liberalism, piety or atheism, willingness to accept orders or intellectual independence. It is, however, inconceivable that there should be a large percentage difference in the genetic types inclined to become Nazis in Germany, France, and Britain, and we must add that research to date on the influence of genes on individual personality has provided very weak or inconsistent results.

However, psychologists have proved that there are major differences among the first, second, and third child in a family. The order of birth has a bearing on precise tendencies that are exclusively cultural and cannot be genetic. In the same way, culture strongly influences all aspects of behavior, as we shall now see.

Culture, a Multifaceted Concept

Encouraged by these considerations, I began to research cultural heredity about twenty-five years ago at Pavia. I developed this project further at Stanford, in close cooperation with a fine mathematician, Marc Feldman, who is also a friend of mine.

From the outset, I felt the need to define culture to my own satisfaction, in order to know what I was meant to be researching. Many years ago, two important U.S. anthropologists, Alfred Kroeber and Clyde Kluckhohn, published a collection of anthropologists' definitions of culture. There were 164 to choose from, but none seemed to hit the nail on the head or in any case improve on the definition I found in the Third New International Webster's: "The total pattern of human behavior and its products embodied in thought, speech, action, and artifacts, and dependent on man's capacity for learning and transmitting knowledge to succeeding generations through the use of tools, language and systems of abstract thought."

Anthropologists have long sought to qualify culture as an exclusively human activity, a presupposition that has undoubtedly conditioned their approach. We now know that many animals have a culture; they make inventions and discoveries and transmit them to their descendants. Events, therefore, have overtaken them—humans do not have a monopoly on culture. But, even if we are not the only animals with a culture, it remains true that we are the animals with most culture; this supremacy is granted by our enormously developed language capability, which undoubtedly is far superior to that of animals and gives us the greatest power in all nature to communicate with one another.

Knowledge cannot be gained without the ability to learn. The basis of culture is the ability to accumulate knowledge, receiving it from previous generations and handing it on to the next so that each new generation need not reinvent the toothbrush, wheel, or integral calculus. Communication between individuals is the cornerstone of any cultural edifice.

Evolution, Complexity, and Progress

Anthropologists have almost always refused to apply the word *evolution* to culture, language, and so on. They prefer to talk of "cultural change," perhaps because they don't want to accept the notion of progress and sanction the classification of peoples as advanced or backward. They are right to be sensitive, although rejection of the expression "cultural evolution" seems excessive.

It may seem natural to conclude that the ethnic groups still clinging to the oldest economic model, hunting and fishing, are most backward, followed by subsistence farmers with their primitive techniques, then more advanced cultivators, and so on, making the economic scale the representative of Progress with a capital P. It is hard to deny that there is a scale of economic progress, but there are other scales too, such as those of beauty, ethics, or happiness (if it can be measured), which probably do not follow economic efficiency at all. To proclaim a culture positive or negative on the basis of its economic importance is not acceptable, and summons the unwelcome ghosts of past racisms.

It should be added that the word *evolution* does not necessarily mean progress. In biological evolution, just as in cultural evolution, the trend is usually toward greater complexity, but this is neither universal nor inevitable, and there are often major exceptions. In any case, greater complexity is not the same thing as progress. Parasites frequently have lost a large number of organs, because the only ones they need to survive are those allowing them to reproduce and penetrate a host, after which they exploit the host's natural mechanisms to procure their food, and so on. Parasites have therefore become less complex in comparison with their distant ancestors, but they are very progressive in terms of adaptation to their present living conditions. Their only weakness is that without their host they perish. But no organism will survive if its environment changes beyond certain limits.

In the cultural sphere, the English language has become much simpler since 1066, but anyone who has to learn it as a foreign language will judge it a gain rather than a loss. To a far greater extent, today's Chinese is also simpler than in earlier times. Those who value the linguistic precision of conjugations and declensions may be concerned that the process of simplification in English and Chinese is a step toward greater ambiguity. The concept of progress is inevitably hard to define either objectively or satisfactorily.

In the long term, there is no doubt that there is progress in biological evolution by those organisms that manage to compete successfully over time, expand numerically and geographically, and perhaps develop useful new functions. In the same way, the efficiency of human language has certainly progressed over millions of years, given that the linguistic tools of our most distant ancestors must have been more primitive.

Fear of accepting the notion of progress is not justified in evolution, whether it be biological or cultural, because of the huge time scales involved. However, it is reasonable for short periods during which the level of change is modest. Changes may sometimes even appear counter-productive, and in any case, it is hard to assess their adaptational significance and future chances of success objectively.

Cultural Transmission

The way we speak, our dress, and our general behavior are a legacy that is received from those preceding us and that constantly alters. Changes in dress style are rapid, much faster than language (although there are inevitable differences between the way the elderly and the young speak, independently of the physical effects of age). This cultural legacy is what makes us recognizably American, Italian, or Pygmy, and is therefore the essence of culture. In what ways is it constant and how does it change? These are fundamental questions, but surprisingly enough none of the books I consulted provided a convincing answer.

Clearly there are forces that render culture constant, or nearly so, and others that induce change. If we compare biology (where genetic knowledge has granted us fuller understanding) and culture (where the mechanisms of conservation and change remain a mystery to us), we find that in biology conservation is permitted by the transmission of hereditary matter from generation to generation, and change is brought about by mutation, the fate of which is decided by necessity (natural selection) and chance (genetic drift).

Can we assume that the same happens in culture? In a certain sense the answer is yes, but the analogy between genetics and culture on an evolutionary level is a little rough and must be qualified. The most important difference is that in biology, hereditary matter is the gene, or chemically speaking, DNA, whereas in culture the hereditary matter is a mass of intangible convictions and knowledge that appear to have no chemical nature, somewhat like the software for that little-known computer that is our brain. When we are able to describe the brain in physical terms, we will find that it is like the content of computer memory (or, more precisely, that it has the same information-processing capacity) but is far more complex; probably knowledge and culture will be described as a collection of states and levels of excitement in nerve cells and their connections.

That culture is more complex than biology will probably also emerge. Human genetic composition is described by a series of three billion nucleotides that constitute the DNA received from the mother, and another three billion, or a little fewer, received from the father. There are, however, at least ten thousand times more cells in our nervous system. The connections, whose role is fundamental because they form the network linking the neurons, are even more numerous. We are as ignorant of the nature of memory and the forces controlling behavior in general, as we were about the nature of genes some fifty years ago when I began to work on the subject.

We know one thing for sure: our motivations appear to be controlled by nerve centers with a known position in the brain, and these determine whether a given sensation is pleasant or unpleasant. These centers undoubtedly influence our behavior in a complex fashion, but the nature of their action is almost entirely unknown. We know of certain substances that probably play a part in determining pleasure, such as endorphins (short for *endogenous morphine*). Through the complex network of nerve fibers in the brain, nearly every sensation and action, as well as every memory, takes on an emotional color that may be positive or negative and is used to orient our behavior. The exact way this happens is still shrouded in mystery, and is one of the great physiological challenges of our time.

Leaving aside what we know of the workings of the brain, the fact remains that we constantly alter our personal knowledge system on the basis of what we learn not only from our own experiences but also, and above all, from the experiences of others—from the body of information, received in the form of orders, advice, or potentially useful ideas. Consciously or subconsciously, we use this knowledge system to guide our behavior.

Who are the "others" transmitting this mass of information? Naturally, that depends on age. In the early years of life, they are our mother, father, siblings, or other permanent members of the household. This circle of contacts gradually broadens to include playmates, their families, teachers, schoolfriends, and nowadays mass media, including the written word, as well. As we grow more independent, anyone can become our mentor or disciple. We never stop learning, but the information absorbed from outside tends to diminish as the years go by, either because we are less receptive or because contacts diminish and our behavior has already been broadly shaped by our experience.

This series of transactions can usefully be called cultural transmission, and it constitutes the vehicle that permits cultural inheritance. Transmission takes place among members of one generation as well as to those of the next. Writing means that information can actually reach us directly from as far back as five thousand years ago, the time of the oldest surviving writings. Archaeology throws light on earlier ages, but the information it provides is reconstructed and necessarily more uncertain and limited in scope.

Vertical Transmission

Modern genetics began when Gregor Mendel, abbot of the Augustinian monastery of Bruenn (now Brno) in Czechoslovakia, discovered his laws of biological heredity, which define the probability of a child inheriting features from one parent or the other, or from more distant ancestors. Mendel's discoveries, published in 1865, were ignored until 1900, when three different researchers rediscovered them or repeated the same experiments. Since then, genetics has taken enormous strides forward, but it still hinges on these rules. Mendel's laws also explain why a biological population remains virtually identical from one generation to the next, unless mutation or other evolutionary factors, such as natural selection and genetic drift, intervene. Even under the influence of these factors, genetic changes in populations are slow, whereas cultural evolution can be very rapid at times.

One part of cultural transmission takes place from parents to children and as such resembles genetic transmission. Theoretical study demonstrates that the evolutionary consequences of this sort of transmission are very similar to those found in biology. Cultural elements transmitted this way behave like genetic features, and are therefore very stable over long periods of time. They are highly conserved.

Our parents' teachings are naturally subject to review as a result of subsequent cultural influences. There is, however, a mechanism that renders some areas of parental teaching particularly effective: humans' greater sensitivity to certain influences during the early years of life. There are critical periods in psychological development during which cultural influences leaves indelible traces; if this influence is missing at the crucial moment, an individual may never develop correctly in the way determined during that phase. This mechanism, known as imprinting is especially strong in animals. A well-known example is that of the duck, which recognizes who its mother is during the first twenty-four hours of life. The first to demonstrate this was Konrad Lorenz, founder of ethology. Lorenz discovered that if the first moving figure a duckling saw was Lorenz or a mechanized puppet, then it adopted Lorenz or the puppet as its mother and began to follow this "mother" around. Such short and precise crucial phases have not yet been observed in humans. But we have seen that people can learn language fully only in early life, and that they cannot learn a foreign language perfectly after puberty.

Other recognized critical periods need fuller analysis. The most interesting to date is the inhibition of the sexual interest between two humans who have been in close relationships long before puberty, or in other words, the unlikelihood of falling for the girl or boy with whom you did your potty training. This theory, first put forward by the sociologist Edward Westermarck at the beginning of the century, is now substantiated by two major examples: In Israel, marriages between children from the same kibbutz—in which all the children grow up together and see their parents for only a few hours a day,—are very rare. In Taiwan, there are still examples of "minor" marriage, the Chinese custom of a family adopting a baby girl deliberately as the future wife of a son born recently. These marriages between adoptive siblings, which are becoming extremely rare, are generally not very successful, although they do allow the mother to mold her future daughter-in-law to her own liking.

There are other examples of results linked to critical phases, even if the duration of each phase has yet to be established and may vary widely. Women with elderly fathers tend to marry older men. It is also said that a man tends to marry a woman who resembles his mother, but the difficulty of objectively assessing such similarities is a major obstacle for researchers. Another example is that people tend to seek environments similar to that of their childhood days. A well-known case is the concentration of Scandinavians in Wisconsin and Minnesota, where the landscape is dotted with large and small lakes, as it is in Sweden. We have done some preliminary research at Stanford on imprinting of this kind: Students were asked how many changes of residence they had before coming to Stanford, and at what ages, and which were their preferences for landscapes—or also, which types of environments they identified with. There was a clear correlation between the environment to which they were exposed in youth, and their preferences, and also between the amount of travelling and residence changes they had been exposed to because of their parents' profession, and their desire to travel or lack of preference for specific landscapes. Our data was not sufficient to pin down with adequate statistical accuracy a particular age interval as being especially sensitive, but the research should have been repeated on older subjects to be more informative. There is here an interesting and important parallel to certain modes of life which involve some degree of nomadism, as is true of African Pygmies and in general hunter-gatherers, or of pastoral nomads, e.g. Bedouins, or gypsies. The latter were probably an Indian caste that migrated to Europe about one thousand years ago, and still speak an Indian language not far from Sanskrit (called *Romani*). All Indian castes are highly specialized professionally, and gypsies were most probably nomadic entertainers. Before radio and television, there were obvious social roles for them, which have now disappeared, creating serious problems for gypsies who have not abandoned nomadism, and for the people among whom they live. The difficulties of convincing nomads to settle are notorious, and are probably due to an imprinting for a mode of life that is acquired very young, is pleasant, and is very difficult to eradicate, like all true imprintings.

The greater impressionability of the young makes parental influence stronger and more incisive, but parents are today losing contact and control, overwhelmed by the deluge of information and activities that absorb the time of both parents and children. Children also have a certain tendency to behave contrary to parental teaching and example; this may create periodic oscillations, perhaps identifiable through dress, which is notoriously cyclical in its variations. Skirt length varies in not very regular cycles from long to short to long and back again, with a secondary cycle for the length of miniskirts. Perhaps the recent revival of anything 1920s, including architecture, completes a sixty year cycle covering roughly two generations, which is what we should expect from a radical rejection of parental taste. This is, however, just amusing speculation for which we lack solid proof.

The fact remains that parents are certainly not the only source of cultural transmission (which is generally positive, and negative only in the event of rejection of parental teaching). Transmission from parents to children and more generally from one generation to the next is known as vertical transmission, because we see it as being passed in a top-down direction; the opposite form of transmission in which age, generation, and kinship don't count, is known as horizontal transmission.

Horizontal Transmission

Horizontal cultural transmission assumes many forms. The simplest is that transferred from one person to another, as in the case of a joke, a recipe, a piece of gossip, or any item of information, whatever its importance. Formally, it is similar to the spread of colds or other contagious conditions, with the difference that jokes or news can be passed over the phone or via the television, but to catch a cold you must come into physical contact with someone who already has one.

Horizontal transmission is in some ways equivalent to an epidemic of contagious disease: news spreads first at accelerating speed, the increase then levels off, and eventually the speed decreases to zero. In particular conditions, the equivalent of endemic disease, where a population contains a certain incidence of an illness for an indefinite period of time, can also occur. Examples of endemic diseases are diphtheria before the availability of vaccination, and tuberculosis prior to antibiotics. In terms of cultural transmission, drug taking provides an example of both phenomena: rapid spread to elements at risk in a population (epidemic), long term persistence of the custom (endemic).

Other major sources of horizontal transmission are teachers, politicians, religious leaders, and other prestigious members of society.

Cultural mentors of this kind may draw tens, hundreds, thousands, or even millions of disciples. The Pope has hundreds of millions of believers who are obliged, at least formally, to obey his precepts. The preachings of the Ayatollahs in support of a holy war have prevailed in much of the Islamic world. A political leader has an enormous audience, and soldiers are obliged to carry out orders given by their superiors. Teachers impart learning to their pupils. Fashion is often promoted by personalities in the public eye. In years gone by, fashion and custom were dictated by royalty and aristocracy. I find the following example delightful. In the closing scene of Shakespeare's *Henry V*, Henry, king of England and conqueror of France, desires to kiss his future bride, Katharine, princess of France. Katharine turns down his request, saying, "It is not a fashion for the maids in France to kiss before they are married." Not in the least put off, Henry kisses her anyway, remarking, "We are the makers of manners, Kate." Today, the speech, clothing, and behavior of well-known actors, sportscasters, and other personalities are widely imitated. When one individual determines the behavior of many people, culture can alter quickly. And, with today's means of communication, if the person transmitting a command or behavioral model has great power or prestige or the new idea is particularly attractive, changes can be extremely rapid. The process is slower where transition is easier for the young than it is for the old, who remain behind and perhaps never accept the innovations. Changes brought in at school that do not influence adults generally take two or three generations to become fully established. The same applies to animals. During research on Japanese macaque monkeys, it was found that a very intelligent young female, Imo, responded to special situations prompted by researchers with various inventions: before eating sandy potatoes, she washed them in the sea, and to separate grains of wheat from sand she floated them on the surface. These new habits were accepted more or less rapidly by other macaque, but in practice they were taken up only by those of her own age or younger.

At the opposite end of the scale of horizontal transmission, we find situations of social pressure where the inverse occurs—many people act on one individual to make him or her accept certain prescripts or new ideas.

The commandments "Thou shalt not steal" and "Thou shalt not kill" are proposed/imposed on all sides generally. Coherent transmission from several sources generally reinforces acceptance of a message. These two fundamental precepts of civilian life and ethics are not necessarily transmitted or accepted universally. "Thou shalt not kill" is often waived for the military or police, and until weapons capable of immobilizing a target at a distance without killing are invented, this situation is unlikely to change. They don't apply in many criminal circles, either, where theft and murder may be the order of the day. The teaching imparted in normal environments and in criminal circles varies radically in this respect. It is possible to switch from one model to the other, even if those brought up under one system or the other normally conform to it for most of their life. This is why it is so important, where possible to keep children from being exposed to the negative influences of families incapable of teaching them normal ethical values.

In most parts of the world, the family is a socially important group. In some cultures the family extends beyond the nucleus of parents and children alone. In polygamous societies, which are the norm in Africa, the extended family is an important reference point for the individual, who can count on the support of vast network of relations. Children from rural areas who go to school in other villages or towns are often looked after for long periods by relations living close to the school. The family group provides sleeping accommodations during journeys. It is a major center of cultural exchange. An example of the enormous influence the family can exert on the individual is the Mafia, which very rarely turns a relation over to the authorities.

The pressure brought to bear by the combined and concentrated influence of many members of a social group on one individual is particularly effective in preventing change, and is therefore a major agent of cultural conservation.

Cultural Mutations

In cases of vertical transmission between generations, or horizontal from a group to an individual, culture clearly tends to remain unaltered and change occurs only with difficulty. The age at which culture is transmitted also has a bearing, because we are more impressionable while we are young; subsequent conversions from upstanding citizen to bandit, vice versa, fascist to communist or Christian to Muslim, or vice versa, are all possible, but are perhaps less likely the older we become. Horizontal cultural transmission from person to person, however, stimulates more rapid change. In top-down transmission, from leader or teacher to followers, cultural change is potentially very rapid, whether the cultural material assumes the form of orders, suggestions, or models to imitate.

For change to occur, some alternative practice must be available. Often the alternative is proposed by the innovation itself, which catches on if it is considered useful or acceptable. In this context, innovation can be defined as an analogy of a "cultural mutation." Its function is similar to that of mutation in biology, except that it generally has some motivation and is not coincidental: it usually represents, in fact, an attempt to resolve a problem. If the attempt proves successful, then the innovation will probably catch on. Sometimes it can prove successful simply because it is innovative or because it is proposed by a popular figure.

One Hundred Ways to Marry

The conventions of marriage vary widely around the world and are an excellent source of material on cultural diversification and the relationship between accepted practice and innovation. How is marriage contracted? Every society has rules and traditions that are often complex and inviolable. In Africa marriage is an important economic transaction. In most cases, the groom or his family purchase the bride from her family, and, for the wealthy, the price to pay may be many cattle and a variety of gifts. The problem is divorce, because the gifts and children remain in the husband's family or return to the wife's depending on which of the two is the "guilty" party. In one region of southwest Africa, the price of a wife has fallen to a pittance; for some reason wives have a low economic value. Among the pygmies, where personal property is minimal, a wife is earned by working for the future in-laws, hunting their meat for months or even years. In other areas pygmies swap sisters, but a sister is not always available, or she may not be to the other pygmy's liking. Perhaps due to Arab influence, in eastern Africa, and to some extent still in Europe, also, the opposite custom prevails and the father of the bride offers a dowry: the groom is purchased, not purchaser.

Why is there such a broad range of customs? The exact sequence of events leading to today's practices would be hard to reconstruct. Perhaps swapping sisters and the idea that parents must be compensated for the loss of their daughter are ancient practices from Africa that have declined over tens of thousands of years until the opposite principle has been reached. In any case, it is thought that the various forms of marriage derive from one another through cultural mutations that have met with greater or lesser success depending on local evolution.

The most immediately comprehensible motives appear to be the economic ones, which are also significant in other forms of marriage, such as monogamy or polygamy. In monogamous cultures either death or divorce must occur before a partner can be changed, but polygamy is more flexible. Polygamy means that a man can have many wives (poly-gyny) or a woman many husbands (polyandry). In Tibet, both kinds of marriage are found in the same villages, and there are also isolated cases of multiple marriages, with more than one husband and more than one wife. Often several sisters marry one man or several brothers one woman, or a mixture. In most Western countries, polygamy is illegal; in other places, it is customary.

In this case, the cultural change may have been the switch from monogamy to polygamy. Hunter-gatherers tend to be monogamous, not least because of the practical difficulties of catching enough meat to support many wives. For the African cultivator, who leaves virtually all field work to the women, the more wives there are, the more food and children (considered a source of wealth). The number of wives a man can have is dictated by the amount of money he can spend. In many areas of Africa, particularly in the tropical forests, farming is still expanding because there is room for new fields, albeit with a low yield, and land is held in common. In Tibet, where society was, until recently, feudal and much less free, polygyny and polyandry are used to resolve the socioeconomic problem of how best to divide land among heirs. If all the male children marry the same woman, or all the sisters take the same husband, the issue is settled without dividing the estate. Elsewhere, primogeniture or similar systems have evolved to cope with the same problem.

Subconscious Motivations

The underlying motivations for other cultural innovations are more complex. The explanation for the low birthrate among hunter-gatherers is very interesting, even if not fully clear to us yet. Hunter-gatherer women generally get pregnant once every four years, and on average, have five children during their fertile life. On average three die before reaching adulthood, and the population maintains demographic equilibrium, with virtually zero growth, thereby preventing overpopulation.

For seminomads who are frequently on the move, there is a major advantage in having a four-year gap between children: one parent can carry the youngest child, the other a few pots and pans and the heavy hunting net, while children over three can walk on their own.

The pygmies keep the birthrate low not to maintain demographic equilibrium or prevent traveling problems, but because a fresh pregnancy would deprive the child of the mother's milk. Children are weaned completely only at the age of three or more, and a sexual taboo has grown up relating to the first three years after a delivery (not all sexual relations are prohibited, only fertilization; and if this happens, the pregnancy can be terminated). Avoiding potential harm to the child is a powerful incentive for Pygmy parents. Long-term breast-feeding has health advantages for the baby since the child's immunity to disease or infection remains high throughout, leading to better chances of survival. On the other hand, breast-feeding diminishes, but does not remove, the risk of a new pregnancy. Otherwise the sexual taboo would not exist.

Today we seek complicated explanations of our forebears' reasons for adopting certain customs, such as male circumcision. Perhaps it was merely a way to resolve the problems of odor and infection due to lack of washing. In many populations it is still an extremely important ceremony and has a social significance above and beyond questions of hygiene. It also reduces the incidence of cancer of the penis, but our forefathers couldn't have known this. Female circumcision (removal of the clitoris) is a cruel, stupid, and dangerous act of mutilation that is supposed to keep women from committing adultery by removing most of their pleasure during sex. It is still commonplace in northern Africa, where circumcised wives are much sought after.

The reasons for adopting certain habits are no doubt complex, and the original motivations may have been forgotten, or perhaps were never consciously known.

Collective Insanity

If we are looking for examples of cultural change, the present is a rich source; they are now so frequent and fast that it is hard to keep track of them. Ours is an age of strong cultural change, the real benefits, reasons, and future of which are hard to identify. The anthropology of modern Western cultures is called sociology; it often uses figures of doubtful reliability to describe phenomena that are fairly obvious without special research. But we are now accustomed to demanding quantitative data for every concept, and, in all fairness, if the data are reliable they should be collected and pondered, even if they describe mundane facts. Despite sociologists' efforts, we are frequently hard put to understand the changes we see around us, or why the changes we would like to see don't come about at all.

Here is one example: variations in birthrate are far more important for the long-term future of the world than are stock exchange reports, and responsible people would like to see the birthrate drop in regions where there is still out-of-control growth. Unfortunately, there isn't much to be optimistic about. A humane and intelligent form of social engineering able to correct the extremely serious errors committed has yet to be invented. In the meantime, the Chinese system of pressuring women into terminating second or subsequent pregnancies may seem hard to accept, but would it really be better if the population of China, which today numbers more than a billion and represents almost one quarter of humanity, were to double every twenty to twenty-five years?

Cultural Change, Genetic Change

Are the changes we see truly cultural? Might they not be genetic?

Even at its fastest, genetic change is extremely slow. One of the fastest major genetic changes is the increase in the number of people able to use lactose, the sugar present in milk. The highest peak registered is 90 percent, in Scandinavia. This level may have been reached over a period of around ten thousand years, starting from an initial incidence of 1 to 2 percent, or maybe lower. The same time lapse may apply to lightening of skin color and generally to the Scandinavians' virtual loss of skin, eye, and hair pigmentation, starting from original colorings that were perhaps similar to the Lebanese of today.

Fast change is unlikely to have genetic basis. One thousand years ago, southern Scandinavia was inhabited by a race of exceptional navigators and colonizers who were also fierce and fearsome warriors: the Vikings. They occupied Scotland, Ireland, Normandy, and Iceland. They reached Greenland and America and even struck at the heart of the Mediterranean. All of this is in marked contrast to their descendants; the Scandinavians of today are calm and mild—Europe's most dedicated pacifists. Some run serious risks and shoulder great responsibility to further the cause of peace in the world. It is difficult to believe this change is genetic, or that all the violent elements have died out in the meantime. Cultural evolution strikes me as a more convincing explanation.

Yet there may have been some genetic change. It is extremely hard to run a genetic analysis of behavioral characteristics. Quantification alone is a minefield, but, above all, personality and behavior are strongly influenced by aspects of the individual's past, which are only rarely identifiable. Both frequently change with age and develop in unpredictable directions. They are often an enigma to the individual him- or herself. There is a component of inner motivation that we rarely want to recognize. How do you measure envy, hypocrisy, anger, or mendacity? Without a measure, valid genetic analysis is difficult. It is complex enough for height, which remains constant—or nearly so—after a certain age and is easy to measure. Blood pressure, which changes frequently, is even harder to study, even though it is relatively easy to measure (and, in fact, medics often let patients take their own readings). We are only now in a position to try to unravel the hereditary nature of disorders such as hypertension, but we find it easier to study the many genes involved in laboratory animals (mice, rats). There are strong indications that the same or a similar set of genes is involved in human hypertension.

Intelligence Quotient

The behavioral characteristic that has been measured most painstakingly is the famous intelligence quotient, or IQ. The IQ test does not measure intelligence itself, which is too elusive and includes numerous facets of different capabilities. It measures the ability to carry out certain numerical, geometric, linguistic, and abstract shape operations. Some have persuaded themselves that it measures only "innate" qualities. Nothing, however, is truly or solely innate in child or adult intelligence. On the contrary, intelligence is the product of personal experience, which is complex and differs from person to person.

Whatever the abilities measured this way, the results are plotted on a fully standardized scale in which 100 or so is the average level of the population, and the variation is such that 95 percent of individuals have an IQ falling in the range 70–130. The scale is calculated to eliminate the effects of age and of gender. If the same person is subjected to a similar but not identical test a short time after the first, the results tend to be much the same. These are all positive indicators that have led some psychologists to believe that the IQ test reveals something significant and useful. In fact, exactly what it measures is not clear—perhaps merely the ability to learn well at school. What is clear is that it doesn't measure only innate qualities, and that it is not "culture-free," or independent of the culture of the country or language in which it is given.

In 1969 a professor at the University of California at Berkeley School of Education, Arthur Jensen, published an article in the highly regarded *Harvard Educational Review* stating that the difference in IQ observed between American whites and blacks—about 15 points, in favor of whites—must be mainly genetic, and therefore irremediable. Initially cautious, he later became adamant about his views, which made him very unpopular in some quarters and equally popular in others. Convinced of the validity of his arguments, Jensen campaigned—in good faith, I believe—for their acceptance. He had the support of the renowned Stanford physicist William Shockley, who had received a joint Nobel Prize for his invention of the transistor. Shockley staged a series of conferences to promote Jensen's convictions, to which he added his own proposal for "social engineering": A monetary reward for black women willing to be sterilized.

In truth, Jensen and Shockley's arguments that the difference between blacks and whites is hereditary were based on extremely indirect evidence and turned out to be ill founded. In a 1970 article in *Scientific American*, and in a book on human population genetics written with Sir Walter Bodmer (another onetime student of R. A. Fisher, then professor at Stanford later at Oxford, and today head of a major British Cancer Research Institute), we demonstrated why those arguments were without basis. On several subsequent occasions, I found myself publicly contesting Jensen and, above all, Shockley.

The poor quality of the schools attended by U.S. black children (particularly during the late 1960s), the problems of motivation for young people brought up in a climate of unbelievable social humiliation, the incidence of extremely difficult family environments with widespread economic deprivation and unemployment, and the lack of adequate parental guidance (often given only by the mother) were and remain enormous, manifest, and well-documented disadvantages. An expert in education should have recognized these as factors influencing the gap in IQ. Only blacks and whites from similar intellectual, economic, and social backgrounds should have been compared, but the considerable—and in part ongoing—segregation of blacks and whites makes this an arduous, and at the that time, even impossible undertaking.

Direct substantiation of our view required extensive, difficult observation. These followed in time, thanks to the efforts of two psychologists, one American and one British, who carried out the only research that could decide the question. Sandra Scarr, the American, noted the high number of black children adopted while still very young by good Minnesota families and compared them with similarly placed white children. Both groups turned out to have similar IQ levels, which were higher than the average for whites. The other researcher, Barbara Tizard, published data from good British orphanages that showed there was no difference between black and white pupils. Properly researched adoption is the only source able to establish whether a feature is, or is not, determined at least in part by biological heredity. This sort of research is hard to conduct, however, and finding children adopted under the conditions required to render observations valid is even harder.

This was how Jensen's theory fell apart. In the meantime, a Harvard psychologist, Robert Herrnstein, put forward a similar idea based on difference in social class rather than ethnic origins. It is well known that IQ, like stature and other physical characteristics, is higher among the upper echelons of society, the gap being even greater than that between blacks and whites, particularly for the extremes of the social scale. Herrnstein argued that this, too, was hereditary, on the basis of the idea that a high IQ is a prerequisite for achieving wealth and social position.

Once again, it was impossible to accurately assess the influence of environment within or outside the family, or the differing quality of schools available to rich and poor children. Again, only adoption could provide an answer. A French study found that the IQ and school grades of working-class children adopted by rich families were essentially the same as those of children born and brought up in wealthy surroundings.

IQ: Heredity or Environment?

This does not mean that heredity has no bearing on intelligence quotient. Research involving adoptions inevitably uses only a small number of cases, and its results are therefore not very precise. There are other methods of establishing the "heritability" of a feature i.e. of estimating the relative role of genes among all factors of individual variation. On their own these methods rarely allow us to distinguish between hereditary and environmental factors, but if we use them together with research on adoption they can help us try to assess the relative importance of the two.

The technique involves measuring the similarities between relations and considering, where possible, numerous different degrees of kinship. The closest relations are identical twins, who account for about one third of all twins (among Caucasoids). Identical twins are literally identical in all genetic features considered, and are very much alike in IQ as well. Other twins are known as fraternal, because they are only as genetically alike as any two brother or sisters. They show greater similarity of IQ than ordinary brothers ot sisters, probably because they share a more similar growth environment.

The resemblance between parents and children can be measured by comparing mother and father separately against each of the children. The degree of similarity between mother and father also must be taken into account, and this tends to be high where IQ is concerned, because we generally marry someone with a similar intellectual outlook. This may be a question of choice, or simply of the opportunities available. This conjugal "match" mechanism poses a number of theoretical problems for data interpretation. Similarities with more distant relations can also be measured, but the results become hazy because the resemblance decreases very rapidly with the degree of kinship.

Are we to assume that the enormous similarity between identical twins is wholly genetic? Growing up so closely and being so alike, identical twins love one another more than ordinary brothers or sisters; they sometimes invent secret languages; they share friends and school and spend a large slice of their early life together. They cannot be said to grow up in independent situations, which is what we would need to reliably estimate the relative effects of heredity and environment.

There is one, somewhat laborious, way of overcoming this obstacle: Go back to adoptions and choose identical twins brought up in different families. They are rare, hard to trace, and often reluctant to be studied. In these surveys, the level of similarity between identical twins growing up in different environments turned out to be lower than identical twins brought up together, but still high. The number of cases surveyed was very limited, and often the families in which the twins were brought up were not completely separate at all. There were various cases of neighboring aunts and uncles each adopting a twin.

A highly renowned psychologist, Sir Cyril Burt, combed British schools for sets of twins adopted by different families, traced several, and published his observations, which pointed to a very strong similarity between them. Here begins a scientific mystery: Many years later, an American psychologist, Leon Kamin, noticed something odd—three comparisons by Burt in three separate projects, presenting observations on a progressively larger number of twins, were identical to the third decimal figure. This seemed a bizarre coincidence, and a British journalist decided to get to the bottom of the matter, discovering, to his great surprise, that one of the cowriters of Burt's works on identical twins had never existed and could only have been invented by Burt. Did this mean that the data on twins were also invented?

By then Burt was dead, and unable to defend himself. The origins of his data were never traced. But how could a scientist, whose work had brought him great fame and even a knighthood, have lost his mind to the point of inventing data? This mystery has never been solved. There is no doubt that twins' IQ is quite similar, while the similarity between parents and adopted children is, in contrast, very low, but the similarities in Burt's data were exaggerated in comparison with the few other studies made. The lesson is that even famous scientists may be so fond of their theories that they will stoop to dishonesty in their defense. Fortunately, all things considered, such occurrences are rare.

There are more recent analyses of identical twins. A full discussion would have to consider the complex interrelations between specific genes affecting traits such as IQ, and environmental conditions. The most satisfactory overall analyses of IQ data estimate that genetics, the developmental environment in the sense of culture (transmitted social environment, that is), and strictly individual factors, in part due to upbringing, have a roughly equal bearing (one third each) on an individual's IQ. This relates, however, only to white Anglo-American populations, and is, in any case, irrelevant to the question of whether the difference in IQ between American blacks and whites is genetic. Regarding the latter, if there is a genetic component, it is minimal, and most of the difference presumably results from the social development environment.

At the end of the 1970s, it was found that the Japanese have an average IQ 11 points higher than the Americans. This is close to the difference between that of white and black Americans which was at the time the discussion was hottest, some twenty-five years ago, around 15 IQ points. This poses a new question: Is the difference between Japanese and U.S. whites genetic or environmental? Oddly enough, I have not heard any suggestion that the difference might be genetic! A debate has begun, however, on the poor quality of U.S. secondary education. This is a positive reaction and will perhaps lead to needed improvements in secondary schooling. If it does something to close the gap between schools in rich and poor areas, it will certainly reduce the difference in IQ between whites and blacks, which contributes heavily to the level of unemployment among black people.

We know that schooling is important and that a good school can increase average IQ. Japanese primary and secondary education is excellent and demands very high levels of both concentration and discipline. In addition, parents are highly committed to their children's education. In Japan, later career prospects rest entirely on the marks earned in primary and secondary school, because they dictate the quality of university a child can apply to, which in turn dictates the quality of the public or private institution that will employ a graduate. In contrast, U.S. families tend to pay little attention to their children's school record, and greet lack of academic success passively, as if they were prepared to accept individual predisposition without attempting to overcome its limitations.

One fact that is perhaps not devoid of significance: From what little I have studied of Japanese writing characters (mostly identical to the Chinese ones, from which they derive), it clearly takes great feats of memory and analytic agility to learn them. The effort required is enormous, but the benefits reaped are probably considerable, too. Your IQ is also, and perhaps above all, determined by how much you are prepared to sweat over intellectual and practical questions that require concentration and analysis.

Two Surveys of Cultural Legacies

Cultural transmission has not yet been analyzed in any depth. This never ceases to amaze me, because it should be the cultural anthropologist's staple diet; it is the factor that preserves cultural legacies through the generations and decides whether a custom or system is to stay or to go, and is particularly useful in the analysis of long-term cultural evolution. It is of less interest to sociologists, whose job is to describe present situations and study immediate changes. In our book, *Cultural Transmission and Evolution* (Princeton University Press, 1981), Marc Feldman and I were concerned above all with providing rigorous proof of our ideas, so we made extensive use of math. We didn't expect to draw much attention with this approach and, fortunately, were proved right as far as anthropologists were concerned. The book was read by economists, however, who are not afraid of numbers. In the end, the real test of our ideas will be how people use them. I have begun a number of projects with some colleagues, the scope of which we hope to widen in the future. I shall briefly describe two of them.

With sociologist Sandy Dornbusch, we asked Stanford University students about their habits, customs, and beliefs and those of their parents and friends. Strong similarities unexpectedly emerged between parents and children on the questions of religion and politics. All the other preferences, habits, and customs (ranging from how much salt they used at meals, to whether they controlled the restaurant check, to superstitions, to times of getting out of bed in the morning) revealed little or no similarity.

We may convert to a new faith later in life, but religion generally is learned in the family. Our data showed that two fundamental aspects of religion are almost exclusively the domain of the mother: the frequency of prayer and the choice of faith when the parents have different religions. In both cases, we seem to be dealing with very early influences, and this would explain their persevering power; religion is usually chosen by the parents well before the child is able to take an active part. In terms of attendance at religious ceremonies, however, the father also wields his influence.

An interesting anecdote in James Boswell's biography of Dr. Samuel Johnson, author of the *Dictionary of the English Language* (1775); reads as follows:

> Her piety [his mother's] was not inferior to her understanding; and to her must be ascribed those early impressions of religion upon her son, from which the world afterwards derived so much benefit. He told me that he remembered distinctly having had the first notice of Heaven, "a place to which good people went," and hell, "a place to which bad people went," communicated to him by her, when a little child in bed with her; and that it might be better fixed in his memory, she sent him to repeat it to Thomas Jackson, their man-servant.

Obviously, Catholicism, Protestantism or other faiths are not transmitted genetically like eye color. Religious faith is the result of cultural influence. If it were genetic, its transmission only by the mother would be even more remarkable. It is true that mitochondria are transmitted this way, but it would indeed be surprising if they were to influence religion.

Cultural transmission of political trends and activities—to which both parents contribute—is almost as strong as religion. Once again, this is probably an early influence, because political discussion frequently occurs in the family. It has been suggested that the structure of the family acts like a microcosm, in which one grows and acquires a predisposition towards specific social ideologies when one later joins the macrocosm formed by adult society. Three types of family structure have been distinguished in France: 1) patriarchal authoritarian (common in the northwest), conditioning children to accepting absolute monarchy and dictatorship; 2) patriarchal benevolent (common in the southwest), favoring acceptance of moderate socialism; 3) strictly nuclear, in which the reciprocal rights of parents and children tend to cease with adulthood of the children. This last family structure, common in the north and east (and also in England) favors migration of young people to areas where there exist opportunities of work and has been especially favorable for the development of the industrial economy and in general economic liberalism. Emmanuel Todd and Herve Le Bras have found correlations with demographic modern and historical data, as well as results of electoral votes in France that agree with this theory.

For some factors, at least, instead of looking at parent-child similarities it is easier and more rewarding to study cultural transmission through the subjects' memories of the teaching of particular activities. With anthropologist Barry Hewlett, we looked at pygmy skills and how they learn the arts of survival in the forest, from hunting to food preparation, child care, dance, and basic social knowledge. In almost 90 percent of cases, the teachers were the parents, or only one parent for tasks distributed on the basis of gender. Very occasionally (virtually only in the case of the crossbow, which is a recent innovation), neighboring farmers taught the skill. Several people from the camp taught social activities. Each pygmy questioned generally remembered quite clearly the time and place in which he or she was taught a particular act or rule.

We stated above that transmission from parents to children, and from an entire social group to its components, are the cultural mechanisms that make innovation hardest to accept. This explains the extremely powerful trend toward cultural conservation among pygmies and other hunter-gatherer groups. The trend is interrupted only when the environment that permits it is destroyed—the forest, in the case of the pygmies.

Culture's capacity for tenacious long-term conservation, when conservation is useful, and rapid change, when change is needed, is a precious adaptive mechanism. It must be acknowledged, however, that sometimes cultural traits prove too tenacious or, at the opposite extreme, too labile, and some extra elasticity or greater stability would be welcome. Even so, humans owe their privileged position to cultural and linguistic skills, which render them particularly effective organisms. We know, however, that this position is very fragile. The civil wars raging around us, the many minorities threatened by various breeds of racism, and fanatical terrorist attacks all remind us how short the step is from heaven to hell.

NINE

Race and Racism

At the New York docks after World War I, my father once saw a job announcement that listed three levels of wage: the top pay was for whites, Italians earned a little less, and blacks the least of all. During that period, after one hundred years of welcoming immigrants from the four corners of the earth (the lines "Give me your tired, your poor, your huddled masses yearning to breathe free, the wretched refuse of your teeming shore" are inscribed on the pedestal of the Statue of Liberty), racism was once again rearing its head. The tragic consequences of the upsurge of racism in Europe are, sadly, all too well known, but even in the United States eugenic proposals were politically very successful. These proposals aimed to improve the human species by encouraging reproduction of the "best" and reducing that of the "unsuitable." In the 1920s, the eugenicists launched a press campaign and put pressure on Congress to pass racist laws and set up immigration quotas to limit entry into the United States of almost all immigrants, except northern Europeans (who, on the whole, were perfectly happy where they were and had no reason to emigrate). Complaining that many immigrants were mentally deficient, they cited IQ results to prove that immigration from southern and eastern Europe was filling the United States with savages. It is true that many were totally illiterate. At the start of the twentieth century, illiteracy in southern Italy, which provided a huge slice of the immigrant population, was still extremely high.

The eugenicists of the time were incompetent scientists but successful politicians, and had their own way over the race laws. One of them, Carl Brigham, afterward (and when it was too late) partially redeemed himself by admitting that the eugenicists' surveys provided not one bit of proof that the differences in behavior and intelligence between immigrant and resident groups were hereditary. In a critical résumé published in 1930, he stated that one of the most pretentious comparisons of race—his own!—was completely without foundation. The eugenicists' scientific leader was C. B. Davenport, founder of the Cold Spring Harbor laboratory near New York. There, he and his assistants dedicated themselves to various human genetics projects, analyzing common features such as the shape of the nose, and eye and hair color. The questions involved were very problematical, and would still be so today, but Davenport didn't realize this, and published many conclusions that were scientifically unacceptable. Genetic research on other organisms was carried out by some exceptionally brilliant minds, however, with the result that many good geneticists in the United States came to the conclusion that human genetic analysis was impossible. Fortunately, the Davenport type of research done at Cold Spring Harbor ended before World War II, and the laboratory later became a marvelous center for the genetic study of organisms other than humans.

Race and Races

To understand racism fully, we need to know what we mean by *race*. The word can be used to indicate the whole of humanity (the human race), but more often than not it indicates just one section. It is often wrongly used as a synonym of *nation* or *people*, a source of considerable confusion. One etymological dictionary defines races as "members of an animal or plant species sharing one or more constant features which distinguish them from other groups within the same species, and which can be transmitted to descendents." The origins of the word are not certain. It appears to date back to the fifteenth century or earlier, and it may derive from the Latin *generatio* or, alternatively, *ratio*, used in the sense of nature or property. Another suggestion is that it comes from "haras," an old French word (still in use) meaning "stud farm."

What matters is that the features are "constant" and "can be transmitted," but we should add "genetically." The meaning of *constant*, however, must be made more precise: does it mean that a feature does not change from person to person or group to group, or that it does not change over time? Both interpretations leave something to be desired. We normally don't know how a biological feature behaves over time, except for skeletal characteristics, which we know have not varied much over the last hundred-thousand years. They are somewhat different, but not much, among extant people. For all other traits, we have to make do with variations between individuals, as they can be observed today. For almost all hereditary features, the differences found between individuals are much greater than those between racial groups. Only rarely do we encounter situations such as that visible for skin color—for which all the individuals of race A are decidedly dark-skinned and all those of race B are fair-skinned.

In short, the level of constancy is not high enough to support the current definition of *race*. Distinguishing race is a complex matter. We have to rely on statistical frequencies for many features in many peoples, never just one. And, to make matters worse, we don't have an answer to the question "How many races exist on earth?"

How Many Races Exist on Earth?

More than a century ago, Charles Darwin clearly identified the most serious difficulties encountered in trying to define human races. The obstacles are so large that we are tempted to abandon the attempt, or to declare from the outset that any and every list is subject to significant limitations. Darwin also points out that the number of races identified varies enormously from researcher to researcher. This remains true today: Recent classifications point to anywhere from three to sixty "races." We could count many more if we wanted, but there doesn't seem to be much purpose. Every classification is equally arbitrary.

The difficulties generally stem from another factor, which Darwin also notes: In passing from one population to another, features often vary continuously and gradually. Even the closest analysis shows that, in genetic frequency maps, discontinuity is rare. Everywhere, change is gradual and constant; in some areas, it just happens a little faster. This can be seen by measuring the genetic variation per mile, which is possible only for areas where the genetic landscape is particularly well-known (which in practice means only for some parts of Europe). It has been suggested that areas where genetic variation is fast and relates to more than one feature should be defined as "genetic boundaries." In Europe these tend to coincide with geographic boundaries, including mountain ranges such as the Alps and Pyrenees, which, however, form only a partial barrier; somewhat wide stretches of sea (the genetic makeup of island populations such as Icelanders and Sardinians is clearly distinct from that of the mainland); major waterways; and, occasionally linguistic barriers, in the absence of geographic or political ones. Regarding linguistic boundaries it is hard to say whether the genetic boundary is the cause or the effect. The genetic boundaries so far mapped out for Europe are, in any case, incomplete and fail to indicate distinct genetic enclaves, the existence of which could contribute to the definition of races; all they do is broadly outline regions in which migration is less frequent. As a result, a slightly more pronounced genetic diversity, which corresponds to small quantitative differences, develops on either side of the boundary.

For all these reasons, it is hard, if not impossible, to classify race and equally hard to answer precise questions such as "Is there such a thing as an Italian race, or a Jewish one?"

The Genetic Geography of Italy

By applying to the Italian population, the genetic mapping technique we have seen for the whole of Europe, Alberto Piazza and his team from Turin University have made an excellent analysis of Italy's genetic landscape. It clearly emerges that the maximum genetic variation is between north and south and that the shift occurs gradually down the length of the Boot. The Italian south is the area where Greek settlements grew up: these stretched from Naples on the west coast (Neapolis, meaning "New City," in ancient Greek, was the name of a settlement founded by Cuma, the oldest Greek settlement in Italy lying slightly to the north), through Reggio Calabria along the sole of the famous boot, and up the Adriatic almost as far as the spur formed by the Gargano Peninsula. The Greeks also colonized eastern Sicily, but not the western part of the island: This is immediately visible in the genetic maps, where there is a clear difference between the western tip and the larger eastern area. The western tip was settled by Phoenicians, Carthaginians, and other people.

Southern Italy was called Magna Grecia, meaning "great Greece," because there were more Greeks there than in Greece itself. Greek was the local language until six or seven centuries ago and remains so in some areas—for example, in a group of nine villages south of Lecce, the most important of which is Calimera ("Good Day" in Greek). The two thousand years of Hellenistic influence left their mark on surnames as well. Throughout the area there are surnames clearly of Greek derivation. The highest level is in the provinces of Reggio Calabria and Messina, where 15 percent of all surnames are Greek in origin.

If the Greeks had a strong influence in the south of Italy, in the north it was the Celts. The Celtic civilization emerged in Austria and Switzerland sometime after 1000 B.C., although its origins may center on an

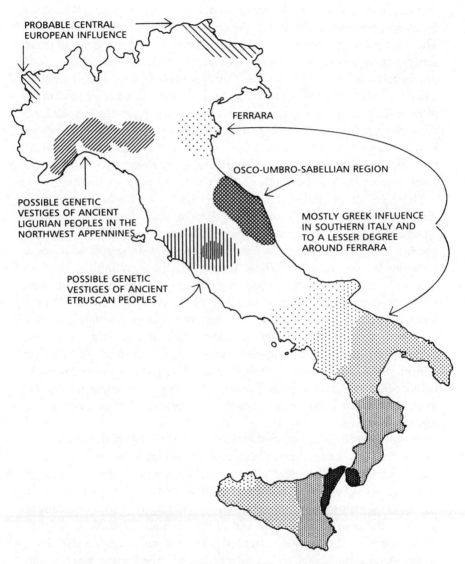

PROBABLE CENTRAL
EUROPEAN INFLUENCE

FERRARA

OSCO-UMBRO-SABELLIAN REGION

POSSIBLE GENETIC
VESTIGES OF ANCIENT
LIGURIAN PEOPLES IN THE
NORTHWEST APPENNINES

MOSTLY GREEK INFLUENCE
IN SOUTHERN ITALY AND
TO A LESSER DEGREE
AROUND FERRARA

POSSIBLE GENETIC
VESTIGES OF ANCIENT
ETRUSCAN PEOPLES

9.1 Geographic map of Italy showing the results of genetic data analyses run by A. Piazza and his team. The different fill-in patterns indicate areas that differ genetically from adjacent ones, probably because the differences that formed long ago have yet to be eroded by exchanges between neighboring villages. These results are very similar to others obtained by G. Zei in a study of the distribution of surnames.

area a little farther east and perhaps to the north in earlier times. This civilization is marked by skill in weaponry and artifacts, and a language that Celtic princes and their armies took to France, Britain, and parts of Spain and northern Italy during the second half of the first millennium. There are some indications that the Celts occupied these regions in quite significant numbers; in addition to donating their languages, which dominated until Roman times, the Celts probably laid the foundations of a large network of villages, as shown by common place-names (including those ending in -ac, like Cognac, in France of course, or Cugnac, and the variations -ago or -asco, leading to Cugnago and Cugnasco in northern Italy). A large number of Celt settlers would justify the genetic similarities observed among the areas of northern Italy, France (particularly the center and east), Austria, southern Germany, and parts of Britain.

There are other traces of ancient peoples in Italy who show up as different from their neighbors on the genetic map. In the Appennines, around Genoa, there are traces of a people who may descend from the ancient Ligurians, a pre-Indo-European people subjugated with some difficulty by the Romans. These traces are characteristically found in mountain areas, where the residents took refuge from invading armies, while the coastal flats were occupied by the newcomers. In the Appennines, between Tuscany and Lazio, there are traces of a people that may date back to the time of the Etruscans; this is the area where the Etruscan civilization first appeared in the early centuries of the first millennium B.C., and subsequently flourished, together with its language, before being supplanted by the Roman civilization. The emperor Claudius took measures to preserve the remains of Etruscan literature, but unfortunately his work has been lost.

Another ancient Italian civilization from the first millennium B.C., has left traces in the area around Ancona. These may be a legacy of the so-called Osco-Umbro-Sabellian civilization. When we have better means of studying the genetics of fossilized human remains, we will be able to check whether these observations correspond to the indications given by the genetic geography and history of these regions. Unfortunately, many of these people burned, rather than buried, their dead, thus making genetic analysis impossible. Luckily, the custom was not universal.

This information is gleaned from maps that condense the geographic variations for many genes, and indicate peoples genetically distinct from their neighbors in the same way that hills and valleys are shown in a contour map. Gianna Zei from Pavia has analyzed a large number of surnames. Surnames generate maps similar to those plotting gene frequencies, but they allow us to make more refined analyses because they are based on a much larger number of people.

A Few European Peoples

On a geographic map, France is almost square shaped. Genetically and historically, each of the four points are different.

The northwestern tip (top left) consists of Brittany, which, as the name suggests, is inhabited largely by people from Great Britain. Even today, the Bretons speak a Celtic language, but this is a secondhand phenomenon, as it were, because when the Anglo-Saxons conquered Britain after the fall of the Roman Empire, many Britons fled to Brittany, taking their language with them and giving their name to the region.

The northeastern corner, near today's Belgium, is genetically closest to central Europe for a number of historical reasons. One of the oldest is the migration of Neolithic cultivators along the river valleys of central Europe and into France; not long ago on the Seine near Paris, a barge used by Neolithic cultivators sixty-five hundred years ago, was discovered. In more recent times, in the fifth to seventh centuries A.D., Germanic tribes from the Cologne region of northern Germany crossed today's Holland and Belgium and settled to the north of Paris. These were the Franks; the country took its name from them, but not its language, which kept its Latin origins.

The history of the peoples of southern France is very different. The south falls into at least two major regions: The east, around Marseilles, which was colonized by the Greeks and still shows some of the genetic results, and the far west, where Basque is still spoken, thanks to the tenacious resistance by an even smaller nucleus of local inhabitants to central government propaganda in favor of the French language.

The Basque-speaking regions were once much more extensive. This can be seen from both place-names and genetics as the French anthropologist and biologist Jacques Ruffié has demonstrated. The territory occupied by the Basques stretches southward beyond the Pyrenees, where there is a far greater number of people still speaking the Basque language. These people probably perpetuate the genes of one of the oldest peoples of Europe, the Cro-Magnon. The three maps show that the region where the best Cro-Magnon artistic work has been found (such as the French Lascaux and Spanish Altamira cave paintings, to mention the

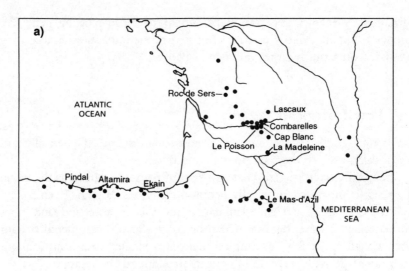

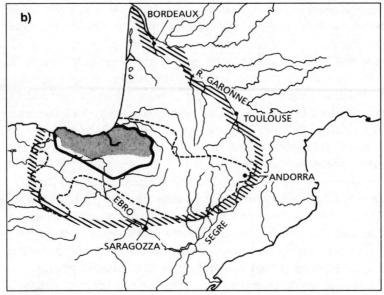

▨ AREA INCLUDING PLACE NAMES CLEARLY OF BASQUE ORIGIN

▦ AREA IN WHICH BASQUE IS STILL SPOKEN

── AREA IN WHICH BASQUE WAS SPOKEN IN THE EIGHTEENTH CENTURY A.D.

---- AREA IN WHICH BASQUE WAS SPOKEN IN THE SIXTH CENTURY A.D.

9.2 Three geographic maps showing the similarity of the geographic distributions of (a) decorated Upper Paleolithic Cro-Magnon cave dwellings, (b) places-names clearly of Basque origin and areas where Basque is still spoken, (c) Bertranpetit's first principal component of western Europe, based on genetic data.

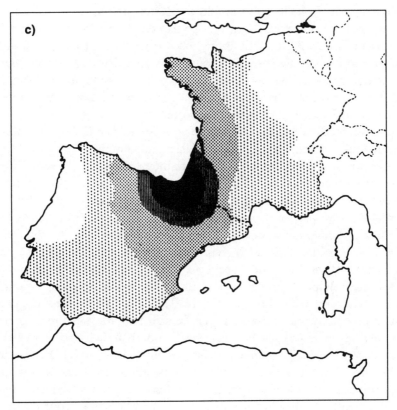

c)

9.2 (*Continued*)

best known) coincides with that inhabited by a genetically homogeneous people straddling the geographic boundary of the Pyrenees.

Is There Such a Thing as the Jewish Race?

This question is particularly interesting, not least because the Jews have been the targets of racist aggression for at least the last two thousand years. For the moment, let us simply consider whether it is right or wrong from a scientific point of view to talk of a Jewish race. Much depends on how detailed we want to be with our definition, bearing in mind that the difficulties posed can be resolved only through complex calculation and better statistical data than we have so far.

The Jews are a heterogeneous people for historical reasons. In the millennia since the two greatest diasporas (literally, dispersions), one following exile from Babylon in 586 B.C., after the Assyrian conquest of Judaea and the other following the Roman emperor Titus's conquest of Jerusalem in A.D. 70, they have reached various parts of Europe. North Africa, and the Middle East. Religious persecution has often forced them to change homes; just one example is their expulsion from Spain in 1492. In recent years many Jews, such as those from Russia, have chosen to return to Israel, if they get the opportunity.

Judaism tends not to be a proselytizing religion. The Jewish communities in Ethiopia and Yemen, however, are probably the result of ancient conversion of the local inhabitants, since they are genetically very different from other Jews and closely resemble the people of their region. Elsewhere, the Jews have preserved not only their religion and traditions, but also at least part of their genetic makeup, a fact demonstrated by the similarities between the various groups. In the course of the diaspora, they have mixed a little with their new and changing neighbors. The incidence of fair hair and blue eyes among the northern European (Ashkenazi) Jews is probably a result of mixed marriages, although natural selection may also play a part. The same is true for the genes determining invisible features. It can be said that the mixture of genes resulting from marriages between Jews and their neighbors over the generations can reach as much as 50 percent (albeit rarely). Bearing in mind the extensive period since initial separation, the level of mix per generation is only a low percentage. The descendants of mixed marriages have shown a greater tendency to lose social and religious contacts with the Jewish community, ceasing to be an integral part of it. This leads to their automatic exclusion from the calculations. The groups of Sephardi Jews, who are today scattered across distant regions, ranging from Spain and Italy to Morocco, Egypt, and Bulgaria, are fairly distinct from each other.

This heterogeneity makes analysis all the more difficult. All we can say is that endogamy (marriage between individuals from the same group) was sufficiently widespread among the forebears of today's Jews for them to continue to have a not insignificant level of genes in common. Therefore, it is not surprising that there is a certain resemblance between Jews of any kind, and also between any Jewish community and the peoples with whom they share their origins, those of the Middle East.

Does this mean we can say there is a Jewish race? If we were to consider just the five races representing the five continents, it is evident that the differences between European Jews and gentiles would be derisory by comparison. Perhaps if we took a large number of races from around the world and compared each with its neighbors, we would find that the Jews resemble their gentile neighbors to the same degree as the northern Italians compared with the southern Italians, or the French from the north with the French from the south. The calculation would be fairly easy, but what would we achieve? It is simpler to say that there is a certain genetic difference between Jews and gentiles, that the average genetic composition of the Jews is not far removed from that of the peoples now living in the nations bordering Israel, and that there is a certain level of heterogeneity among Jews deriving from mixed marriages after the diasporas, but that an underlying "family feeling" still remains.

It is not easy to compare genetics and culture, but the overall impression is that the forces uniting the Jews so strongly are cultural rather than genetic. The Jewish people have preserved their identity mainly through their traditions, on which religion probably has a significant, but perhaps not all-important, influence. The concept of race is so vague that we could call the Jews a race—or, better, a group of races—only if we were prepared to define thousands of races, each of which differs only very slightly from the others.

The idea of race in the human species serves no purpose. The structure of human populations is extremely complex and changes from area to area; there are always nuances deriving from continual migration across and within the borders of every nation, which make clear distinctions impossible.

Racism and Racial Purity

Let us take another definition of race, similar but not identical to the first: A group of people united by common origins, who to some extent are similar genetically, in terms of inherited biological features. They may also have conserved cultural identity and share traditions, language, and political unity, or may have lost any or all of these. Cultural identity is generally less stable than its genetic counterpart; for this reason, we will take into account only genetic identity in defining a race. This definition is unsatisfactory, because it is very difficult to use; as we said, there could be a few, a few dozen, hundreds, or even thousands of races on earth, and all classifications are arbitrary. Still, some definition is helpful if we want to speak about racism.

Racism has many origins and definitions, but we know that racists often worry about racial "purity." Let us dispense with this aspect first: There are no pure races, and if we tried to create one, the results would be most uninviting. It doesn't take much to prove this. In any genetic system, we register a high degree of what is known as *polymorphism*, or genetic variation, meaning that a gene is found in different forms. This is true of almost any gene, and also of any community be it a nation, town, or village. For example, the proportions of A, B, and O genes fluctuate from village to village, town to town, and nation to nation. In every microcosm, we find a genetic composition comparable to that of the larger group, albeit a little different. ABO is remarkable in being exceptional: A and B are usually missing among most Native Americans. We can look at the wealthy or poor sections of society, whites or blacks, but the phenomenon remains the same. What is the point in talking about the "purity" of a race, when every group of people, no matter how small, is variable? If we look at a different continent, the genetic proportions are a little different, but every microcosm still tends to reflect the macrocosm. Genetic purity simply does not exist in human populations.

Genetic purity could be attained, to a certain extent, through a breeding program using very close relations—by marrying brothers and sisters, fathers and daughters, and so on. Unions of this kind are banned in most human communities. We would need, however, to run the program for twenty to thirty generations, and even then we would not have obtained a perfectly pure group in which all the genetic variation has been eliminated. This is possible, in an imperfect way, with animals or plants, but we know that one of the usual consequences is very high sterility, which makes it difficult to keep such breeds alive.

The reality is entirely the opposite: to guarantee normal fertility and health, marriage between close relations must be avoided, or at least kept within limits. In general, mixed marriages, including those between people of very different origins, create a more robust line of descendants. There is absolutely no known biological disadvantage to interracial marriage.

Racism

Racism is the conviction that one race is biologically superior to the others. That is what underlies racists' concern for the "purity" of the race, they do not want this superiority to diminish or cease. But we know that no race is pure. Therefore, to think about conserving purity is absurd. That nearly everyone born in certain areas of Scandinavia is fair-haired, or nearly all the Arabs are dark, by no means signifies that the same "purity" exists for other features. The high incidence of features, like light skin color and hair, and a few others, is in all likelihood simply the outcome of natural selection due to climate over a period of time. For any other genes, the fair-haired Swedes are as variable, and "impure," as any non-Scandinavian. In the same way, the selection of dogs, horses, or other animals to achieve homogeneity of external characteristics, such as color of fur or hair, body or limb shape, or sharp sense of smell in dogs and pigs, running speed in horses or greyhounds, or herding ability in sheepdogs, has little or no influence on the vast individual differences existing for all other features. The breeder who oversteps the mark with these homogenization techniques by breeding very close relations, in the hope of attaining a "purer" result, risks losing the race altogether through low fertility or generally depressed viability.

Although today we are fully convinced that pure and perfect races cannot exist, in the past the false ideal of racial purity has been the cornerstone of many invalid theories, and some have had a significant influence on history. Of these, the ideas of Count Joseph Arthur de Gobineau in the nineteenth century are worth mentioning. He began his career as secretary of Alexis de Tocqueville, the famous French essayist and politician. Gobineau was a traveled diplomat and the author of many books. In his essay *The Inequality of Human Races* (1853–1855), he argued that the supreme race was the Germans, purported to be the purest decendants of a legendary people, the Aryans. He suggested that civilizations decline because of ethnic interbreeding, which reduces the vitality of a race and corrupts it. In a sense, Gobineau was a precursor of the ideas that inspired Wagner, Nietzsche, and Hitler himself. There were other people with similar ideas, but he was the most influential.

Racism is, however, older than these ideologies and probably as old as humanity. Usually, we think our own "race" is the top one (if we take race to indicate a social group), independently of whether the justification we provide is biological (we are better looking or smarter than the others) or sociocultural (our life is more pleasant). People often make no attempt to separate biology and culture, but to consider them one and the same thing is a mistake. In Gobineau's time, there was no attempt at distinguishing between the two.

In the more distant past, the Greeks held all foreigners in contempt; they called them barbarians, meaning stutterers, because they didn't know Greek. Probably every ethnic group has developed a certain group pride, which can interfere with objective comparison. As a racist, Gobineau was unusual in that instead of singling out his own people as the elect, he opted for a group of foreigners, the Germans. It is true, however, that the northeast French and many aristocrats can, rightly or wrongly, boast decendancy from the Franks, the Germanic barbarians who invaded northeast France after the decline of the Roman Empire. The English also can boast a Germanic pedigree, through the Anglo-Saxon invasions. One Englishman, Houston Stewart Chamberlain, who married the composer Wagner's daughter, became a great admirer of the Germans and champion of the false idea of Aryan supremacy.

Apart from anything else, the idea is a recent invention. The term Aryan was first used in the study of linguistics during the nineteenth century to define the Indo-Iranian group of languages. The Indo-European root, ari or *arya*, means nobleman or leader. Hitler fell in love with the word, but perhaps he would have adopted another had he been aware of its real origins: the Indians are without doubt far more different from the Nordic blonds than, for example, the Jews, whom he chose to hate above all others.

We all know (and those who don't are requested to read their history books a little more carefully or get better ones) the turn that German racism took when Hitler became leader of the country. We hoped the lesson had been learned, but the newspapers of the last years and months increasingly prove how wretchedly easy it is to forget the past and repeat the same mistakes. Across the whole of Europe, there is an enormous resurgence of racist sentiment; it is even springing up in regions where it used to be rare or nonexistent. Weren't the six million Jews wiped out in Nazi concentration camps enough? There are even those who maintain that the concentration camps never existed. How is this possible? How can one propose or accept in a civilized country the idea of "ethnic cleansing?" Are we to conclude that racism is an incurable social disease, destined to torment us forever?

The Origins of Alleged Biological Superiority

The Europe of modern times has developed great political and economic power: France and Britain have experienced centuries of glory and grandeur, the vestiges of which remain despite the undoubtedly drastic decline. For three centuries, Spain was a wealthy, conquering nation. In other parts of the world, there have been other empires, lasting at most, a few centuries and sometimes only a few seasons.

The constant changes in power throughout history show how shifting it is and how difficult it is to keep for very long. Usually, success and power go hand in hand. The euphoric sensation of belonging to the world's top nation, or at least one of them, with all the associated advantages, easily induces us to believe that this supremacy is objective, innate, destined to last, whereas in reality it is the result of clever and lucky policies that may well prove ephemeral. History shows that these fortunate periods don't last—indeed, that they are destined to melt away, sometimes rapidly. With success gone, where does that leave the alleged superiority? It no longer has one sound argument in its favor. It is unthinkable that in the few generations it takes for even the greatest civilization to decline, the genetic code of a people can change, perhaps as a result of racial inbreeding—in particular with blacks or Orientals—as Gobineau thought.

This alleged biological superiority of the successful nations of the day, which no one can prove, stems from the confusion of culture or civilization with genetic makeup, and nation with population. Gobineau's arguments make unedifying reading, since without producing one scrap of data, he claims to show that the decline of every civilization is caused by the mixing of races, and that all humanity's progress has been achieved through the efforts of a few Aryans. Despite this, his erroneous racial theories duped a large slice of European intelligentsia for almost a century. Naturally enough, it was easy to convince the direct beneficiaries, the Germans, whose conviction lasted longest, and also had the worst consequences. There is not much point in blaming Gobineau; in his wake many others expounded similar ideas, some of them quite independently. In any case, the roots of racism go far deeper than the views of one aristocratic intellectual.

The Pathogenesis of Racism

Various elements combine to make racism a form of deviancy, which is not at all unexpected. Racism is just one manifestation of a broader syndrome, xenophobia, the fear or hatred of foreigners, and more generally of those who are different. Misogyny falls into the latter category, but we may need to coin a new word—misoandry—for the opposite phobia, women who hate men. Other phobias are directed against homosexuals, priests, blacks, Jews, and the rest.

Social group is very important in determining a person's lifestyle, and it seem reasonable to think that there is a strong urge to feel and act in a way that will gain approval from one's own circle, in order to obtain support and also to provide it where necessary. However, the fact that it seems reasonable for this urge to exist by no means *proves* that it does. Hard proof is difficult to find in the behavioral sciences. But let us assume there really is an urge, an innate tendency, to consider the group we belong to as an entity, which we will call *Us*, as opposed to other groups to which we do not belong. Which we shall call *Them*.

If we accept this notion, we must also accept that the definition of *Us* varies according to the circumstances. *Us* may be the family, or perhaps the family minus some members we don't feel deserve our support and trust. The family is certainly the first *Us* group everyone comes into contact with, except those not lucky enough to have a family, or not a good one. As we progress through life and make other social contacts, other *Us* groups become important. These may be playmates, school-friends, and other members of the school community, or organizations we gradually join. Later on, there are working colleagues and various clubs or associations, each of which creates a new circle that forms a new *Us* group. A number of these *Us* groups may conflict with one another; for example, our family may not want us to see certain friends or schoolmates. This scenario can be of great significance for an individual. Many of the *Us* groups created to spend our time more pleasurably would be particularly interesting to look at from an anthropological point of view: "Our" football, basketball, or baseball team, and so on can become very meaningful to us. Especially in towns with one or more home teams, it is almost essential to choose and actively support just one of them.

The different *Us* groups that influence so much of our life are enormously important as sources of pleasure and anxiety, jealousy, anger, and guilt. They determine a sense of loyalty and belonging that includes various forms of patriotism, but also a wide range of provincial rivalries. Their importance in everyday life suggests that there is an innate tendency to construct these *Us* groups, which are an extension of the self and help us build a protective barrier around ourselves.

The tendency may be stronger in some than in others. If one or another *Us* group takes on a special importance because it helps replace another (the family, for example) that, rightly or wrongly, fails to satisfy needs or desires, the outcome may be acute conflict.

This alone is not enough to explain racism. There are other major determining components. One is the power of prejudice, which can reach the proportions of a serious neurosis. For some unknown reason, we frequently come up against such half-baked, rigidly held views, at times in highly intelligent people, that we feel can be classified only as neuroses. A classic example: A number of individuals, fortunately the great minority, pass their time analyzing the supposed misdemeanors of the Jews. The most extreme manifestation is found in Hitler's semitophobia. The Jews become these neurotics' target because they are socially successful, and, despite numerous persecutions, have always bounced back and some of them subsequently achieved positions of great standing in the arts, sciences, finance—in short, in all areas where they have been given access.

Jealousy and envy are frequent causes of racism, but so is overestimation of the worth of oneself and one's group, and contempt for others. Many of the Pygmies' neighbors genuinely regard them as animals. Racism is not the sole prerogative of Europeans and Americans: It turns up everywhere. Many years ago, in a police station in the central African Republic, I happened to read a circular by Jean Bedel Bokassa, president of the Republic before he turned out to be a megalomaniac and proclaimed himself emperor. The circular said that everyone should be respected as an individual, regardless of the group to which he or she belongs. It repeated that the commandment "zo we zo" (in the Republic's official language, "a human is a human") should always be followed. The phrase was probably not his own; it is more likely to have been the work of the Republic's first president, Barthelemy Boganda, a close relative of Bokassa and a man of great worth, who died prematurely in a plane accident. It is interesting that the Central African Republic was for many years under the political control of a fairly small tribe, the Ngbaka, who had close contacts with the Pygmies. Having known them well over a long period of time, the Ngbaka recognized and appreciated the Pygmies numerous qualities.

Contempt of individuals belonging to a social group poorer than one's own (like Pygmies, who are always at the bottom of the socioeconomic scale) is another source of racism. This can be exacerbated by the unhappiness of the social group that feels superior. Unhappiness, whatever its source, generally stimulates the search for a scapegoat, who is always somebody weaker. In the United States the ethnic groups who arrived last were always the poorest. They suffered from the racism of the groups who had arrived before them, who had had the time to adjust somewhat, enough to feel superior. Although by far not the last arrivals, today the greatest victims are African-Americans. They have been in the United States for two or three centuries and slavery was declared incompatible with the constitution in 1865, but it was only in 1954 that public schools were desegregated, and only in 1964 that a congressional act declared segregation in the private sector unconstitutional as well. But even if everything goes well, it takes three generations to even out the initial handicap.

It is not surprising, therefore, that there is still a large economic gap. We cannot hope that the African-Americans will manage to close it with the same relative ease as immigrants of European origin. The main barrier, which is also the major test of racial equality, remains almost intact: The acceptability of mixed marriage. While marriage between whites and Americans of Asiatic, Amerind, and Polynesian derivation (in the Hawaiian islands) is more common, marriage between African and white Americans has increased very little since the first steps toward equal rights were taken. As the geneticist Curt Stern pointed out many years ago, if skin pigmentation were irrelevant in selecting a marriage partner, the color gap between black and white would disappear with two to three generations. The fact is that the cultural and economic differences are still too strong, and awareness of skin color is too ingrained, for the situation to change with satisfactory speed.

Racism is a chronic disease that one cannot hope to suppress rapidly or easily. But the frequency of terrorist racist actions, whether organized by governments, secret or not-so-secret societies, and gang warfare has become so high that countries should take strong measures to prevent them. There are two major therapies. One is preventive and cannot be expected to produce rapid results, but is absolutely fundamental: It is education at all levels. The other is political, juridical, social strong medicine to curb the explosion of violence and fanaticism that we witness. Parliaments should give appropriate, very strong laws; judges should mete out very strong punishments, and police should be up to the needs of very strong enforcement of such laws. Helping the social integration of such immigrants, by offering them good opportunities of learning useful skills and of becoming acquainted with the social structures of the host country, is also of the utmost importance.

There is a hope, but it is for a relatively distant future. The increase in communication and migration is eroding the separations among people, and also among races. We are still very far from the "melting pot" which, at the beginning of the century, it was believed the United States would be; however, migration favors both genetic and cultural melting. And, if genetic melting is increasing only very slowly, cultural melting is proceeding faster. Whether this will be effective in reducing racism remains to be seen; but it is clear that the final weapon against racism is education, along with other social policies that can lend a strong hand in the same direction.

TEN

Evolution and Progress

Since his appearance one hundred thousand years ago, *Homo sapiens sapiens* has changed and diversified into the groups found on the earth today. Modern humans have developed advanced powers of communication, but still have a long way to go before they use this skill well and learn to live peacefully together. It would be very difficult to change our hardware, or our genetic makeup. It is much easier to try improving our software, or culture. Social progress has been made: Slavery was widely practiced almost everywhere for thousands of years, but in the twentieth century it has virtually disappeared as a legalized institution. The inhuman working hours of the 1800s and the habit of sending children down narrow mine shafts in Britain have both gone, too. At least in the West, the terrible exploitation of humans by their fellow humans is no longer tolerated by law.

In reality, every day the press carries reports of acts and events as awful as those of the world wars, which we hoped we had put behind us. Technological progress has been enormous, but this is not necessarily positive, since it has also had numerous highly damaging side effects, which ignorance, laziness, and greed have prevented us from recognizing until too late. Obvious examples are the accumulation of colossal trash piles, contamination of the air and sea, the systematic destruction of the forests that allow us to breathe on the earth, the indiscriminate sacking of nonrenewable energy sources, which have taken millions of years to create and just one century to consume.

Technological progress is in itself neutral: everything depends on whether it is applied well or badly. We see the possible extremes especially in nuclear power. Nuclear power may represent the solution to the energy problem, which is going to assume extremely worrysome proportions. However, there need to be fundamental improvements in the peaceful applications of nuclear energy to avoid pollution and the risk of serious accidents. On the other hand, its military applications now present a very serious threat to the very survival of humanity; there is a small, but not indifferent, chance that vast regions—or even the entire planet—along with all its inhabitants could be destroyed by one deranged tin-pot dictator, or the error of some high-ranking bureaucrat.

How Will Human Genes Change?

The forces of evolution have been altered radically by the developments of the last ten thousand years. The number of people living on the planet has increased over a thousandfold since agriculture began. As a result, the effects of genetic drift are now much more modest, and we could almost say shelved. From this point of view, it is very unlikely that the existing groups will continue to diversify.

Some types of natural selection have also been shelved. Until a few centuries ago, 50 percent of children died before puberty, most of them in the first year of life. Nowadays very few die: in advanced countries infant mortality is often lower than 1 percent. Although the levels have not dropped to the same extent, infant mortality is now generally much lower in the third world, too. What has hardly changed at all is the birthrate. As a consequence, population is increasing at a staggering pace, and in some countries, particularly in Africa, South America, and South Asia, it may double over the next twenty years.

For natural selection to work, some have to die where others survive, and some have to die more easily than others. Plummeting infant mortality has almost eliminated the effects of natural selection due to differences in mortality. There are still a number of genetic disorders we cannot cure. The ones that would have the worst consequences are never even seen because the mother miscarries, often without knowing, in early pregnancy. An ever-increasing number of genetic diseases can be avoided before birth, usually by terminating the pregnancy.

Various religions have chosen to see abortion as a crime. For the greater part of humanity, however, abortion is a sometimes necessary and acceptable practice, despite the distress it causes. If the Chinese government did not encourage abortion as a means of controlling birthrate, the Chinese population, which currently accounts for one quarter of all humanity, would grow at an explosive rate and would soon need to drive out the rest of humanity—except perhaps for India and a few other nations where all attempts to halt population growth have so far proved unsuccessful—just to have the room to live. Chinese law forbids couples to have more than one child, or two in certain cases. This is a severe and undoubtedly painful imposition, but it is unavoidable and deserves the gratitude of the rest of the world.

Governments' failure to change reproductive habits is understandable, because they are extremely difficult to control. The position of religious authorities that refuse to help humanity in this necessary crusade is unjustifiable. Condemnation of abortion and almost all methods of contraception in the name of the right to life, is burying one's head in the sand. It is a refusal to see the monstrous extermination awaiting humanity in little more than a generation, as a result of excessive expansion—extermination through famine, epidemic, and war, the three great forces redressing demographic equilibrium since the beginning of life on earth. These are the weapons that Providence, in whom believers place such faith, will unleash when the human species outstrips the dimensions permitted by its environment.

Curiously Italy, the world's most Catholic country and seat of the pope, one of the most intransigent religious leaders of all regarding contraception, also has the lowest birthrate. The dictates of the Catholic church obviously are not followed to the letter. The use of abortion as a means of controlling birthrate is considered regrettable, but it is important to be able to correct mistakes. Furthermore, abortion not only is a means of birth control, but is currently our only way to limit certain major hereditary diseases.

Natural selection works not only through mortality but also through fertility. Variations in fertility can provoke significant changes. In the last one hundred years (and earlier in some regions), the wealthy classes have reproduced less, to the extent that some people, mistaking wealth for intelligence, have seen the specter of a decline in average intelligence. The phenomenon is transitory, and stems from the fact that the decline in birthrate has taken place first among the upper, and particularly professional, classes, before spreading down the social scale right to nonspecialized laborers.

Fertility varies greatly from family to family: this is partly biological and can in some ways be selective, but the influence of culture is also very strong. For example, in Britain and the United States, there are relatively few Catholics, who, however, probably obey the dictates of the church more conscientiously than other peoples such as Italians, Spaniards, and French, who have much higher percentages of Roman Catholics, but some of the world's lowest birth rates. Even on a superficial level, there is a noticeable difference between the silence that reigns during mass in Anglo-Saxon countries, and the hubbub and general lack of attention surrounding celebration of mass in Italy.

Natural selection through fertility may continue to operate as an evolutionary factor, but it may have a leveling influence on all features when it favors heterozygotes (those who receive different forms of the same genes from their mother and father). The result is that extreme types tend to be passed over in favor of intermediate ones. Natural selection is significantly changing the world's population in another sense, because it is changing the numerical relationship between the various races, however we define them. The very high birthrate in Africa, Brazil, India, and many other nations in the southern hemisphere is inevitably changing the overall composition of the human species. News of this will make white racists blanch, but it has its consolations, particularly in terms of the consumption of resources.

Europeans and North Americans consume huge quantities of energy and resources to produce all their goods and services, and have no alternative but to go on doing so if they wish to maintain the lifestyle to which they are accustomed. If they reproduced at the rate of African nations, the world would rapidly be drained of raw materials, or alternatively be forced to adopt a much poorer lifestyle. It is possible that the drop in the number of births in economically advanced nations reflects the need to avoid this collapse in consumption and all it implies in terms of standard of living.

The real problem will be the demand for energy, food, and raw materials in the Southern Hemisphere, which would be hard enough to satisfy even without population growth. In these countries, however, population now doubles every two decades, making improvements in the standard of living almost impossible. If anything, we should expect to see serious socioeconomic disintegration, or the intervention of nature and its methods (famine, plague, and war). Isn't that already happening? It takes a good deal of optimism not to become very alarmed.

All these considerations relate to sociopolitical facts. As far as genetics is concerned the average human will evolve very little. The most significant fact will be the shifting ratio between races. Individual migratory exchanges will also inevitably become more frequent, as will interracial marriage, which is in no way bad news, even if it would have thrown Gobineau and company into a tizzy.

Eugenics

The term *eugenics* was introduced by Francis Galton, a cousin of Charles Darwin and pioneer of the study of human genetics, in 1883 to express the concept of making improvements in humanity's genetic stock.

Eugenics has two forms: positive and negative. Negative eugenics is the elimination of physical and mental defects; positive eugenics increases the frequency of more desirable qualities. In many states and a number of Scandinavian countries there are laws for compulsory sterilization of people suffering from some conditions, particularly certain forms of mental deficiency. For various reasons, these laws have never been applied to any great degree. Apart from anything else, individuals with the most serious conditions are often naturally sterile and, in any case, sterilization is an inefficient and ethically dubious way of eliminating a genetic disease. For many inherited defects, we still don't know enough about the genes responsible or how they are transmitted. It would take the sterilization of an enormous number of people to eliminate other diseases, because of their nature.

For example, cystic fibrosis is a genetic disorder with an incidence of one out of every two thousand births in Europe. Real understanding of its mechanisms is only now being achieved; to wipe it out we would have to sterilize one person in twenty, because that is the frequency of "healthy carriers" (now easily detected in most cases). Cystic fibrosis leads, or led, to almost certain death in childhood through a series of lung and intestinal problems; even now doctors often diagnose it because of a characteristic peculiar to the affected children, that of exuding high levels of salt (sodium chloride) when perspiring. A German proverb says that the child with a salty skin will not live long. The prognosis is somewhat better these days: average life expectancy has reached twenty to thirty years, if the children receive intensive, and costly care. This is just one genetic disorder, albeit one of the most frequent. There are thousands of others, almost all of which are rarer and for the most part inadequately understood. The final number is impressive: if the compulsory sterilization were extended to all the carriers of hereditary disorders, perhaps there would be no one left who would be allowed to reproduce.

Disorders like cystic fibrosis are called "recessive," from a Latin word meaning "to remain hidden." In fact, it remains hidden in healthy carriers and emerges only when two healthy carriers start a family, and then only in an average of one out of four of their children. Other diseases described as "dominant" show up in the carrier, sometimes very

250 THE GREAT HUMAN DIASPORAS

late in life. One terrible dominant disorder, Huntington's chorea, which causes death through the slow and inexorable deterioration of the mind and nervous system, usually appears about the age of forty, with death on average at fifty. By that age, any children have usually been born, and these have a fifty-fifty chance of developing the condition. In this case, sterilization would work. The chance of correct diagnosis is practically 100 percent, but it will always be someone's thankless duty to inform a healthy human being that he or she will slowly die in a psychiatric hospital from a terrible disease. Children of a parent with Huntington's chorea are reluctant to be tested and most (nearly 90 percent) don't. But it is hard not to sympathize with individuals exposed to this risk. If the result is positive, life can become hell.

A serious recessive form of anemia, thalassemia, which we have described at length, has practically disappeared from areas where it was once common (in Cyprus, as well as in Sardinia and Ferrara in Italy, for example), thanks to antenatal testing and termination of the pregnancy where the condition is detected. The decline, however, cannot be seen as a form of negative eugenics. The disease used to be frequent in both areas and is so feared that newly married couples willingly come forward to be tested, and are prepared to terminate a pregnancy if the test result is positive. As a consequence, thalassemic babies are very rare these days. By rights, the local Catholic priests ought to discourage the termination of pregnancy, but luckily they have little influence or perhaps take pity on the family that would otherwise have the terrible task of rearing a sick child needing constant blood transfusions (and therefore exposed to the risk of AIDS or hepatitis), whose life expectancy is in any case limited. Marrow transplant now provides effective therapy, but a matching donor is not always easy to find; the family, or society, has to pay bills of hundreds of thousands of dollars. The increasing cost of medical care will force economically developed countries to spend an ever greater proportion of their revenues on health if they want to provide adequate care. The question is, would it be economically possible to make the costly operations required if all the testable embryos potentially affected by the hereditary disorders were to be born into the world? The answer is almost certainly no.

Antenatal diagnosis of hereditary disorders and termination of pregnancy are not examples of eugenics. They are prophylactic procedures. Avoiding the birth of sick children does not reduce the future frequency of the disorder. On an emotional level, however, the solution is acceptable for the vast majority of parents. It is much less complex and risky than allowing sick children requiring lifelong medical care to come into the world. Also despite well-documented religious resistance to abortion, it is hard to understand why we should not avoid the enormous suffering caused for both patient and close relations, leaving parents with the thankless task of one day explaining to their own child that he or she is dying of an incurable disease, not to mention the serious economic consequences for family and society.

Negative eugenics may not seem to be a new technique. The Romans threw severely deformed or incurably sick newborn babies from the Tarpeian Rock near the Forum. The Spartans had similar customs. Many primitive societies practice infanticide for the same reasons: It is an act that is harder to accept than abortion, but a primitive society does not have the means to keep very sick babies alive for long. However, it would be wrong to equate these practices to negative eugenics; in fact, the incidence of genetic disorders in a population is practically unaffected by them. Strictly speaking, negative eugenics proposes avoidance not only of the birth, but also of the conception of individuals with physical or mental handicaps. The truth is that even if we wanted to adopt them, negative eugenic programs are not yet possible in practice. All we can do for the moment is monitor the most frequent and best-known genetic disorders and prevent children suffering from the worst of them from being born.

I think we need to be clear on one point: we believe no one should ever feel obliged to terminate a pregnancy in order to avoid the birth of an incurably sick child; but parents can and must be counseled on the implications of genetic disorders. If there is a chance that a child may be born with a serious condition, parents have a right to know, and to know that there are solutions, regardless of whether the doctor concerned disapproves of abortion on religious grounds.

And positive eugenics? The idea of improving the human species is not such a strange one, considering that domesticated animals and plants have shown constant genetic improvements over thousands of years. Some, such as the maize cob (see Figure 6.8), have made huge leaps: from a length of about half an inch eight thousand years ago, the average maize cob has gradually grown to today's size. Cattle and sheep have been selected to obtain higher-quality, more profitable milk, meat, or wool. The best idea of the power that selective breeding can have is perhaps provided by dogs, a species where selection has created an unparalleled range of breeds.

But would it be so great to have perfect waiters and waitresses, secretaries, soldiers, and so on? A tyrant might like the idea—some kings have tried to implement positive eugenics (it is said that the eighteenth century Prussian king Frederick the Great had his Pomeranian grenadiers marry the prettiest girls)—but it is profoundly contrary to human dignity and also human needs. The results of individual cases of crossbreeding are always very uncertain. Isadora Duncan, the celebrated American dancer, suggested to George Bernard Shaw that they should marry, and create a child with his intelligence and her looks. Shaw refused on the famous grounds that the child might end up inheriting his looks and her intelligence!

We need to maintain the existing biological diversity, because we do not know what future challenges lie in store for us, and in particular what infectious diseases. A terrible new germ, the AIDS virus, has recently appeared. We know next to nothing of the possible variations in individual predisposition to this disease, although we know they normally exist for all infectious diseases. If we were to eliminate this variation, or accidentally started to reproduce individuals prone to the disease, it could mean the end of humanity.

If we wantonly selected a particularly promising-looking type, we could profoundly change the human species. When artificial insemination was organized in Denmark, five bulls were chosen to provide sperm for the new generation of Danish cattle. Unbeknownst to the breeders, one was affected by a genetically linked heart disease, which as a result became widespread throughout the country's bovine population.

A precise proposal for positive eugenics was put forward by one of America's most brilliant geneticists, Herman J. Muller (who also discovered the mutagenic capabilities of X rays for which he was awarded a Nobel Prize). He suggested using the sperm of exceptionally intelligent and gifted men to artificially inseminate women volunteers (a procedure known as eutelegenesis). Politically, he was a communist, and he spent some time in Russia in the interwar years. It is said that he tried unsuccessfully to interest Stalin in his positive eugenics program. Dissatisfied with the results of his Soviet experience, he struck all the communist leaders off his list of chosen few whose genes should be propagated.

In recent times, an American industrialist financed an initiative in eutelegenesis, setting up a sperm bank for famous personalities and offering access to interested women. Many Nobel Prize winners refused to donate; Linus Pauling said he felt that the natural method was better. One Nobel Prize winner who was interested was William Shockley, the physicist whose interest in eugenics led him to spend considerable time on the subject of a possible genetic basis to the difference in IQ between blacks and whites. On an earlier occasion, Shockley has been contacted by a married woman seeking artificial insemination. The couple were childless, because the husband was sterile, and both agreed on insemination by Shockley. The lawyers of the two parties met to draw up a contract. One lawyer raised a problem: who would bear the maintenance costs if the child happened to be affected by a serious congenital defect? This was obviously a possible scenario. On the strength of this doubt, the agreement foundered and the donation of sperm never took place.

The new techniques available pose this and other, similar questions of bioethics. Above and beyond the legal and ethical problems, artificial insemination using famous men as fathers has provoked curiosity and some mirth. There is something vainglorious and even laughable in the noble gesture of an illustrious person donating his sperm for the betterment of humanity. Surely more modest donors would be preferable, just in case modesty is hereditary?

Many factors make the application of eugenics inadvisable. It seems almost self-evidently desirable to produce individuals who are good, intelligent, brave, and so on. In reality though, we don't know to what extent these psychological traits are genetically controlled, or how they work. A consideration made by a colleague provides a measure of our ignorance. Schizophrenia is a major social disease because of its high frequency (1 to 2 percent of births), and the fact that it causes fits of insanity that can have very serious social consequences. We know there is a hereditary component to the condition, even if we have not yet managed to identify it. It seems, however, that among many schizophrenics and their relations there is also a certain flair for the arts and general creativity. A similar correlation is found for depressive syndromes. Were we to suppress the genes responsible for these diseases, we might run the risk of losing much art, theater, and literature.

Genetic Engineering

In the labs at Stanford University and San Francisco in 1973 a sensational experiment was completed: a segment of bacterial DNA was inserted by chemical linkage into a minichromosome capable of entering and reproducing in another bacterium. In this way, the extraneous DNA was shown to be capable of functioning. This made it possible to transfer segments of DNA from one organism to a very different one, in a totally new manner of creating hybrids, and opened the way to previously unheard of uses. One of the first medical applications was to make a bacterium produce the human hormone, insulin, used to treat diabetes. Many others, such as the growth hormone, interferon, TPA, and various growth factors followed in its wake.

Genetic engineering is the construction of new organisms in which a section of DNA has been artificially modified, or has been replaced by a section taken from a different organism or perhaps made by synthesis. The above examples are obtained by inserting into bacteria a given piece of the human genome that has been modified to function and reproduce a large quantity of the substance desired. The range of potential applications in every field is enormous, from the treatment of hereditary diseases to the improvement of crops or animals.

When these applications still lay far in the future, the pioneers of DNA research and its genetic engineering applications asked themselves if their work could have grave or unforeseen consequences. Soon it was suggested that a moratorium be placed on this kind of study, and very severe controls established to avoid the release of genetically engineered bacteria that might produce an uncontrollable plague. When subsequent developments showed that these fears were unfounded, many of the precautionary measures were dropped. The process of genetic engineering is not "tampering with nature," as it may perhaps appear. Nature itself provides examples of mechanisms that operate in a similar way. All the methods used to extract and recombine DNA and insert it into chromosomes use specific enzymes very common in nature. The fact that some scientists of great standing voiced exaggerated concern had undesirable effects on the minds of some who began to have futuristic nightmare fantasies and project genetic engineering as the work of the devil. After a few years of unbridled fear in which experiments were subject to extremely stringent controls, a more normal situation finally emerged.

It is my belief, however, that it was good for scientists to face the possibility that they might have been doing great harm so early on, and publicly. It cost them much in terms of limitations to the scope of their work, limitations that were above all self-imposed. It is hard to find comparable examples of sense of responsibility in any other scientific field.

In general, however, we cannot hope to predict all the negative side effects of the new industrial applications. Was it foreseeable that the piston-engine would have helped the growth of gigantic cities, in which the air was ever harder to breathe? Or, to cite an example where the link between cause and effect is much more direct, that asbestos would turn out to be so damaging to lung tissue? There are two alternatives: one is to halt all scientific and technical progress, which would be extremely perilous because new problems requiring rapid reaction and remedies are constantly emerging (AIDS is one example yet to be resolved); the other is to start social engineering in earnest, to monitor new problems as they emerge, study the best way to treat and prevent them, and quickly pass suitable legal measures. The fact that we are not yet in a position to do this doesn't mean that it cannot be done.

We can always imagine some evil dictator setting up research to clone armies of highly trained and disciplined soldiers, and other forms of servants needed to take over the world. Of course, to do so he or she would have to be able to genetically control psychological traits to perfection, and there is no proof that this is possible. The level of control probably could never be perfect enough to allow the project to be implemented. In any case, we are far from having the knowledge required to create such a nightmare.

Modification of our genetic makeup through genetic engineering is not yet possible, and won't be for some time to come. Everything tried so far, and to the tiniest degree achieved, has involved *non*germinal cells, which are also known as *somatic* cells; these changes cannot be transmitted to later generations. Humanity doesn't yet have the technical knowledge or, for that matter, the moral wisdom to undertake genetic improvement itself. Modifying somatic cells seems on the other hand, permissible, and desirable to avoid major disorders.

With present progress in artificial insemination, however, it is becoming possible to check whether there are signs of known faulty genes in the cells of a developing embryo. Inserting genes into it is another matter and potentially dangerous for the progeny. It should not be done, in my view.

The Human Genome Project

The Human Genome Project has been discussed for several years. Now, at last, the project is beginning to take shape. The objective is to map the entire genome; that is, all the genes present in the chromosomes. This simply means writing down the sequence of three billion letters A, T, C, and G, (indicating the four nucleotides) that compose the twenty-three chromosomes. With sixty spaces per line, fifty lines per page, and three hundred pages per volume, we are talking of an encyclopedia of more than three thousand volumes, forming a respectable-size library that will offer, it must be admitted, terribly dreary reading—but also a description of the whole genetic blueprint of a man or woman.

So far, much rougher maps of chromosomes have been drawn up, filling only a few pages, but they form the basis of the future, complete map. One major objection raised is that the work will occupy many teams of fine researchers for several years on an initiative providing little intellectual stimulus. In reality, the problems posed will be highly exacting at times and take a great deal of effort to resolve. Another major objection is the high level of funding—some three billion dollars—required. This means tying up resources otherwise available for other biological and scientific projects, unless there is an overall increase in research funding. A third possible objection is that not all of the genome provides useful information, since there are parasitic segments, sometimes defined as "selfish." Until recently, these were thought to serve little purpose, but we now know that they can become harmful if they undergo certain mutations.

What can we expect to gain from this study? We will learn the sequence of nucleotides of DNA in the structural genes that build proteins, the fundamental molecules for the cellular metabolism. The sequence of these genes allows us to deduce the structure of the proteins, from which we can hope to infer their function. Many disorders are caused by mutations in these genes. Today, understanding the position and nature of the gene that determines a certain pathology takes years of work by a laboratory team of twenty or so researchers. When the whole genome is known, it will be much easier to identify the genes responsible, and this may open up new opportunities for therapy as well.

Other sequences are fundamentally important in understanding how genes, cells, and our entire organism work. There are those that switch the genes' function "on" and "off," and regulate their productivity. There are undoubtedly many other structures, functions, and properties we know nothing of and can only now begin to identify. As the new sequences become known, the huge theoretical task of interpretation will begin—a task without equal in the history of biology. Three billion nucleotides are a gargantuan amount of data, enough to try the mettle of the most powerful computer. But we do not know what computers will be available when the study of the human genome will be complete.

Human Genome Diversity

The Human Genome Project has an inherent flaw: The three billion nucleotides, or the three thousand volumes needed to write them out, relate only to one genome, to half of an individual's DNA, that is. This is because they describe the genetic code received from only one parent; that received from the other parent is in principle the same, but somewhat different. We wouldn't need to fill another three thousand volumes to describe this second edition of the genome, since most of it is identical to the first. The fact is, however, that the second half would include several new elements, and neither of the two is necessarily better or more significant. Both are equally qualified to represent humanity. If we take a new individual, we will find further variations, and so on with each new individual. To what extent, then, is it worth describing the human genome? How many people do we have to analyze before we have done the job properly?

We cannot answer this question fully, because we have only a vague idea of the degree of variation awaiting us. We think that the average variation is such that the difference between the DNA we receive, say, from our two parents is about one every 300 or 400 nucleotides. This means the difference between the parental DNAs is, on average, around ten million nucleotides. However, we also know that this level can be higher or lower, according to the section of genome analyzed. For very important genes, which cannot change without serious or tragic consequences for the individual, the variations are more limited, perhaps about one nucleotide out of every thousand, or even less.

The Human Genome Project would therefore be incomplete, and perhaps fail in a number of its most important aims, were we to take just one—or, more precisely, half of one—individual. As it stands, it is a formidable undertaking, so the study of individual variation should be planned with the maximum economy. It is unthinkable to fully analyze the sequence of nucleotides in a hundred or thousand individuals. An intelligent program, however, can hope to cover most of the significant individual variations, with a very modest expense, less than 1 percent of the whole budget. A group of colleagues and I have launched a proposal for such a program, called Human Genome Diversity.

A pilot project had already started on a much more modest scale—in 1984 when I returned to the Pygmies I knew in the Central African Republic to gather samples for a collection of blood cell cultures providing the DNA needed to study the donors' genome. Blood contains large numbers of red corpuscles and one thousand times fewer white corpuscles. Only the white cells are able to reproduce because they have a nucleus, which the red corpuscles have lost. A small fraction of the white cells, known as B-lymphocytes, are capable of reproducing indefinitely if we treat them in a test tube with a special virus, known as Epstein-Barr. Cultures of this kind can provide us with any amount of DNA that is practically identical to that found in the other cells of the same individual.

These cells are quite fragile, and to grow cultures in the lab we must use fairly fresh blood. It is difficult to chill the samples without sacrificing vitality; the best solution is to tuck them in your pocket, as it were, and take them to a lab for proper treatment as soon as possible. The populations whose DNA we most want to preserve with this technique often live far from airports, and this poses serious practical difficulties. With the help of a colleague, in 1985 I took blood samples from the Ituri pygmies of northeast Zaire. As evening came, we boarded a small American mission plane, which fortunately had been able to land on a nearby airstrip. This way we were able to get to the mission the same night and sleep there. In the morning, one of the mission's larger planes, which runs a shuttle service on certain days, took us to Nairobi; from there we flew to Europe, and the next morning we caught an onward flight to the United States. A fair proportion of our samples made it in good condition to the lab of my colleague, Kenneth Kidd, professor in genetics at Yale. Kenneth's wife, Judy was able to grow many cultures successfully and was one of the first people to use the procedure of transformation as it is normally known. In much the same way a year earlier, we had already transformed samples from another group of Pygmies in the Central African Republic. This community lived a few hours' drive from an international airport (which meant the logistical problems were less severe).

Together with the Kidds and members of the lab team at Stanford, we have so far been able to transform samples for an average of some forty members of fifteen populations around the world; these are giving us an initial idea of the world's genetic variation by looking directly at DNA. All the other data we discuss in this book have been gathered over the last fifty years using different methods that do not directly analyze DNA. The results they furnish are reliable, but for many reasons they are less complete and satisfying than those obtainable working directly on DNA.

This served as a pilot for the later, much more ambitious Human Genome Diversity program. The president of HUGO, (the Human Genome Project international organization) at that time was Sir Walter Bodmer, the director of London's famous Cancer Research Institute. Walter and I have worked together on a number of research projects and two major books, one describing the genetics of human populations and the other a textbook on genetics and human evolution. As president of the project, Walter, who like me recognizes the need to study individual variation, set up a committee to study human genome diversity with me as chairman. Initially, Allan Wilson, the author of the brilliant studies on mitochondrial DNA, commonly—but imprecisely—known as African Eve, also was a member. Unfortunately, when the committee was formed in 1991, he was already suffering from acute leukemia, and a bone marrow transplant failed to save his life. The committee today is composed of twelve geneticists, from around the world. Many more scientists are becoming interested in the venture, especially in Third World countries where most of the populations to be sampled are found. The project has started to function in Europe and Asia.

The Importance of a Multidisciplinary Approach

One challenging and exciting aspect of the program is its multidisciplinary nature. This sort of study is impossible without the help of people from many different disciplines, ranging from anthropology (whether it be physical, cultural, or ethnographic), linguistics, archaeology, and history, to human geography, economics, and demography. Special subjects such as toponymy (the study of place names), the study of surnames, rock painting, and probably numerous others also have a major contribution to make. We need generalists as well as specialists from disparate fields. The British novelist and scientist C. P. Snow remarks in *The Two Cultures* on the broad gap between sciences and the humanities (represented by history, literature, and the arts). In this project, we have to bridge the gap. Researchers from totally different fields can work together easily; even very closely, if it is in the interests of both and both are willing to learn the major notions and terms from the other's sphere.

Scientific terminology is useful in specialist work, but it can stifle the spread of information and comprehension between different disciplines. It is important to reduce jargon to an absolute minimum, in order not to frighten off potential helpers. Terminology's only useful purpose is to make communication between experts in the same field faster and more precise. It is pointless to apply its sometimes cryptic and narrow forms elsewhere; on occasion it is abused in much the same way as doctors used to fall back on Latin terms and abbreviations when they wanted to prevent the patient from understanding what they were saying.

The concepts dealt with in most disciplines are generally not so obscure that they are beyond our grasp, if expressed in simple terms. To achieve good interdisciplinary cooperation, terminology must be kept to a minimum and re-explained every time it is used, or at least every so often. The chance to work in new fields is a great opportunity, because years of dedication to increasingly specialized work in one subject can become stultifying. The introduction of an additional focus of interest for part of the time can be truly revitalizing.

There is another basic problem, already mentioned, that only an interdisciplinary approach will help solve. Like all historical work, the analysis of evolution has a deep basic flaw, at least for those who are used to dealing with natural sciences: it lacks the support provided by experimentation. History cannot be repeated exactly as it happened nor, above all, at will. What is done is done, and many details that might help explain certain developments are lost forever.

We may profoundly admire an incisive, detailed, and apparently thoroughly convincing historical analysis, but we know that historical conclusions always contain an element of uncertainty. In some case, analysis of an evolutionary process can be helped by computer simulations that, by reproducing parts of real events, tell us if alternative theories provide equally valid explanations to a phenomenon. The doubt remains, however, that we may have overlooked one or more significant factors when programming the simulation. In addition, in every set of real events (and likewise in every computer simulation) the element of chance renders every interpretation of the fact observed open to doubt.

Experimental research in chemistry, physics, or biology has the comforting advantage that experiments can usually be repeated and the conditions changed at will. That the conclusions reached are much more solid than those possible in studies of evolution therefore seems inevitable. This is generally true, but even experimental work can involve subtle interpretations that generate doubts, and render results that are not completely reproducible. Only mathematics provides the certainty of attaining indisputable conclusions.

Experiments are subject to error and a result is generally considered credible only when it has been obtained independently by at least two researchers. In all fields there are numerous examples of errors and even a few deliberate deceits, that remain undiscovered for a long time. When dealing with abstract theoretical constructions based on complex experiments, the element of doubt may always remain.

The Proof of the Pudding

Theories that still seem rash even after lab results have indicated they are sound, may become credible only when applied. It is perhaps hard to believe the geneticist claiming to describe the sequence of nucleotides that regulates the rate at which tomatoes rot, but if a chemist uses this description to produce artificial DNA, inserts the DNA into the genetic code of a strain of tomatoes, and the tomatoes start to rot much more slowly—perhaps in a way never seen before in nature—then it becomes harder to reject the initial theory. Genetic engineering has now provided numerous ways to slow down the rotting process in tomatoes; this means that they can be harvested later, when they are riper and juicier. Some of these products will soon be launched for the first time and face the test of consumer reaction. Hundreds of other vegetables could follow. However, hundreds of people also are terrified of genetic engineering, and they are very active politically.

In research on evolution, a multidisciplinary attack offers the most important insurance against mistakes. The question of how Europe came to be peopled could be tackled from many angles, of which genetics is but one (in my opinion a particularly incisive one). By comparing genetic data with that from other sources, we can see if a notion holds up under different lights and also discover that certain hypotheses make a lot of sense. The ideal occurs when all the methods and sources of information that are used point to one correct interpretation, but this is hard to achieve; many possible approaches may have little to offer on a specific problem. Moreover specialists in a field are often not prepared to bear in mind conclusions reached by tackling their problems from a different standpoint or using methods that differ—from theirs. It may take a long time to bring them around, and sometimes one senses that only the grave will part them from their prejudices.

It is inevitable that in a lifetime of research a scientist will make many mistakes. A researcher must recognize that mistakes are possible, although they are always regrettable when they happen. It is important for scientific researchers to arm themselves with a good dose of humility. Historical research is even riskier, and open to interminable diatribes. Despite this, it can also be more satisfying, because the obstacles it presents encourage great mental agility.

The major factor influencing the length of arguments among researchers is uncertainty, which tends to be greater in biology than it is in physics, and in physics greater than in mathematics. In anthropology the uncertainty reaches a peak, engendering a great deal of discussion and criticism. Some U.S. schools of cultural anthropology train the young professional anthropologists to enter the ring of scientific activity as if it were a dogfight.

The Study of Humanity

Leaving aside questions concerning the sociology of science, the problems involving humanity are fascinating. Studying our origins and past helps us to understand ourselves. Much of our life depends on our cultural background—as well as on another fundamental factor, our genetic structure.

Even illnesses are to a great degree expressions of human culture and history. A certain number of illnesses are the direct result of our biological constitution, or that of the germs and parasites that attack us from all sides. Many are the result of human technology. The switch from hunting and gathering to agriculture has brought with it various disorders. Some, such as intolerance to milk, gluten, or a nutritional deficiency found in developing countries (kwashiorkor), are now comparatively rare; others, such as hypertension, arteriosclerosis and some tumors, which depend on excessive intake of fatty foods, have become increasingly common. Prior to the development of a certain type of economy based on agriculture and livestock, they were rare. The animals eaten by hunter-gatherers lead far too active a life to accumulate fat deposits. An enormous range of diseases is linked to our cultural environment.

Health is said to be the major basis of happiness. However, other components, such as the work we are best suited to and leisure activities we pursue, also have a role to play. These, too, depend on our biological and cultural history, and change from person to person. To develop our personality harmoniously, we need to study and respect individual variation, be it cultural or biological.

In this sense, understanding history and evolution has much to contribute, because there is evidently a high degree of practical as well as intellectual interest in these fields. Humanity deeply needs to understand itself better and learn to exploit its cultural inheritance in far better ways.

Epilogue

Agriculture and livestock farming: These inventions from ten thousand years ago are applications of primitive genetics. Reinvented in this century, this science has made giant leaps forward and has revealed the nature of life itself. It has given us extraordinary power to modify living organisms, even if few of the potential applications have so far been developed, and mainly in the field of medicine. However, it is clear we are on the threshold of a new era.

As always, these potential applications can have harmful, or even wantonly evil, side effects. It is up to us to guide them. Achievements in crop and animal farming have been extremely important to humanity and helped us overcome one crisis ten thousand years ago. But they have also prepared the way for others. For example, the indiscriminate grazing of goats, sheep, and cattle can destroy dry environments, particularly fragile ones, and quickly provoke irreversible changes. The dessication of the Sahara was, and in part continues to be, a result of one such mistake. The once highly fertile region of Mesopotamia has largely become desert because irrigation for agriculture has led to an excessive increase in the salt level in the soil. These disastrous consequences could not have been foreseen at the time. The use of the horse for military purposes, however, was clearly bound to encourage fierce wars. The introduction of cavalry more than three thousand years ago brought about a revolution of war techniques comparable to that of the invention of firearms a few centuries back.

The medical applications of genetics are directed at the treatment and prevention of hereditary diseases, and are nothing to fear. We can prevent the birth of children with some of the most serious and widespread conditions, and we could eliminate almost entirely the birth of children with major genetic diseases. So far, this has meant terminating the pregnancy, but in the future less invasive methods may become possible.

The intense suffering caused by hereditary diseases for both those affected and their close relations is already largely avoidable, and can be completely so. It is therefore incomprehensible that most theologians of the Roman Catholic church and their less subtle, but even more rigid, fundamentalist counterparts in other religions should be so eager to condemn the parents of future sick children either to suffer this way or to never have children of their own. The punishment is visited on the children as well as the parents, and this is profoundly unjust, because in a better world a child would have the *right* to be born healthy. Theological concepts deny them this right, even when human knowledge has made it possible.

Some enthusiastic eugenicists would like to create a better breed, not of dogs, horses, or sheep, but of humans themselves. To choose the most important qualities may look easy; but it is not simply a matter of copying the plans for improving animals and plant species (known as "artificial selection"), and applying them to humans. In the case of animals and plants, it is relatively simple to decide which features to improve; in humans, the major ones are difficult to observe or measure. Furthermore, the project would be laden with enormous moral problems regarding the violation of personal freedom, dignity, and fundamental rights; in any case, even if we wanted to try, we lack most of the necessary knowledge to proceed.

The most ardent eugenicists are usually spurred on by at least two forms of fanaticism: the conviction that genetics is paramount in deciding our strengths and weaknesses, and the belief that it is fundamentally impossible to change inborn conditions. They are in part so-called do-gooders, but they seriously risk leading us down the road to hell, even if the path would be paved with good intentions. It is much better to seek moderate, farsighted corrective action, aimed at eliminating the most serious defects.

The idea that what is genetic cannot be corrected is often erroneous, and progress is foreseeable in this direction. Perhaps with time, we will be able to cure some of the forms now considered hopeless; but letting the ones that are now partially curable increase freely in number could be very expensive in both human and economic terms. Medicine does progress, but at its own rate.

Some other social issues are less costly to resolve than health problems and are today even more pressing. It is evident that our intellectual, moral, and social upbringing is deficient and must be improved. Salvage operations are urgently needed to solve high unemployment, train people for new jobs, and create new specializations. Although the choice of a career is one of the most important decisions in life, it is usually made on the basis of inadequate information and with very limited chances of finding a job really suited to our personality. We need to ascribe proper weight to the important step of entering the workforce. Work continues to occupy the largest part of our waking life, and it is important to make it more satisfying than it generally is today. The ideal job is the one you enjoy more than any leisure activity. Naturally, it is hard, and perhaps impossible, for everyone to achieve such a position, but we must come closer than we are at the moment. A good beginning would be special teaching programs, providing precise career information early, and the opportunity to seriously try out more than one job and even change careers after some years, if necessary. If people could choose their activity freely, making the most of their personal talents, we would have greater hopes of creating a secure and pleasant existence that not only encourages a variety of tastes, inclinations, and activities but also minimizes the injustice and cruelty repeatedly perpetrated around the world. History shows that civilizations flourish where variety of expression and very disparate contributions are exploited; they decline when intolerance and the inability to interact with those who are different prevail.

Political, religious, and racial persecution daily fill the world stage. Racial persecution, which today is particularly acute, is even more monstrous than religious or political persecution, because if we cannot avoid violence from groups stronger than ourselves, at least in case of danger we can change or hide our politics or faith; but not our race.

We all recognize the importance of knowing the past, because history holds the key to phenomena and expressions of human life that would otherwise appear incomprehensible. The biological history of humankind is that of its evolution, and cultural history is an integral part that has both influenced it and been influenced by it. The two must become inseparable if we wish to avoid our heartrending ocean of suffering. The animal part of our nature, often lacking in restraint, is responsible for many of these excesses, but our cultural history should teach us how to avoid them.

Where can we find a guide in our doubt and consolation for misfortune when it strikes? The traditional source is religion, an activity that appears to distinguish humans from animals. We naturally turn to the faith we were taught from birth in a given region or social environment. Many rely on religion for comfort, but this does not always seem to create a better environment. On the contrary, each religion's conviction that it is the sole repository of the truth has generated the worst conflicts of all. The respect felt for the religion that we happen to have inherited can hardly come from its dogmas, which are for the most part a useless, if not actively harmful, superstructure, nor from its history, which is checkered with violence and contradictions. On the contrary, it comes from the belief that the majority of religions share a common ethical substrate, also shared by many philosophies, which is simple, easy to grasp, and able to resolve most everyday moral issues. Many so-called pagan peoples accept the same principles.

At the end of the day, problems of individual moral behavior are relatively easy to solve on the basis of very few ethical principles, which more or less all of us do learn. Decisions of importance concerning a social group, or, worse, the whole of human society may present more difficult moral problems, and in addition, unusual technical ones, arising from the need to predict the long term consequences of an action. It is the responsibility of every human community to predict and prevent potentially dangerous uses of its economy, science, or technology, by developing prospective and preventive methods and early warning systems. This alone will allow us to achieve a more peaceful existence, and to enjoy the wealth and variety offered by the balanced development of human potential.

 Postscript

Déjà vu

When I moved to Stanford University, about twenty-five years ago, a furor had been started by an article written by Arthur Jensen, a professor of education at Berkeley, and published in the *Harvard Educational Review*[1]. As summarized in Chapter 8, it claimed that the average intelligence of blacks in the U.S., as measured by IQ, was much lower than that of whites; and, the "heritability" of IQ being very high, the black/white difference was of genetic nature. Under the assumption that inherited behavior cannot be changed, the take-home message was that there was little hope, if any, of correcting this unfortunate gap.

Even prior to Jensen's publication, William Shockley, a physicist at Stanford who had won a Nobel prize as coinventor of the transistor, had been very vociferous on the same question. He felt it was necessary to prevent the disaster of a reduction in national intelligence, which he believed was impending. His solution was radically eugenic. At least at the time, in the U.S. blacks had a higher birth rate than whites and in his view, a continuation of this trend could only lower the average IQ of the nation. His solution was to give a bonus of $5000 to black women who agreed to be sterilized.

There was a clear lack of genetic understanding in the thinking of Jensen and Shockley which led them to serious errors. As the only human geneticist at Stanford, I had to react. Shockley was touring the country to give lectures on the subject and I repeatedly had to debate him. Walter Bodmer and I had just written a book on human population genetics[2] in which we had considered the same problem at length. But the book had not yet been published, so we put together a summary of the relevant

parts of it in an article that appeared in *Scientific American* in 1970[3].
Another session of fireworks, though less impressive than that started by
Jensen's paper, was the initiative of Richard Herrnstein[4], professor of psy-
chology at Harvard, who was more concerned with the differences in IQ
among socioeconomic classes. These are greater than those between
blacks and whites, being at least 2 to 3 times as much.

After a while, the furor calmed down. Not much happened on this
front for the next twenty-five years, except for a few important papers on
the IQs of black children adopted into white families and the books mentioned briefly in Chapter 8. More than twenty years later, the book *The
Bell Curve*[5] exploded like a bomb, spreading what are essentially the same
old ideas to a large public. Is it a concidence that the book appeared at a
time when Republicans seem to be taking power, and want people to lis-
ten to some scientific motivations of an agenda rather similar to the one
they suggest? Unquestionably, Herrnstein and Murray's recommenda-
tions have considerable similarity with those cherished by more extreme
conservatives: reduce to the bone, if not cancel completely, social services,
affirmative action, welfare programs, federal intervention in education,
etc. Their dream is to go back to a pre-Roosevelt era.

The book is ably written, with a technique that reminds me of that used
by Shockley, who deluged the public with graphs difficult to understand
for most people, but designed to convince his audience of the validity of
his eugenic proposals. Throughout their book, the authors of *The Bell
Curve* are in fact always using graphs to convince you, for instance, that
the IQ of parents are more important than their socioeconomic status
(SES) in keeping away a number of social ills, from poverty to illegitima-
cy to crime. In this way they claim to have proved the social importance
of IQ. But it is well known that one of the most difficult problems of
socioeconomic analysis is to dissect the relative importance of causal factors on the basis of correlation coefficients.

"Correlation" is a statistical technique for measuring similarity (or its
opposite: dissimilarity). But even a mediocre statistician knows that the
price of butter and that of cosmetics are likely to go up or down in parallel, when we compare different times, or different places. In the case of
different times this is because of similar effects of inflation on all or most
prices. When considering different places, e.g., different states, the correlation is determined by overall variation in the general cost of life in those
different states. Similarity of the two prices does not imply that the price
of butter causes that of cosmetics, or vice versa. I cannot avoid feeling

scared by the arrogance with which the authors always know what interpretation of data is right or wrong, what is good for the country, and what we should do. Of course, politicians must always pretend they are right. But I have never found scientists who always know the sure answers. It is rare that the authors concede ignorance or inability to decide on the basis of available evidence, unless it goes against their beliefs. But most of their conclusions are really based on insufficient evidence. In these socioeconomic problems the possibility of good experimentation is extremely limited or totally absent, making it extremely easy to make mistakes. It is not as in normal scientific endeavors in which it is possible to carry out proper experiments, varying at will the weight of potentially important factors and testing rigorously their effects.

Some of the early mistakes by Jensen and Shockley were not repeated, at least superficially, by Herrnstein and Murray. For instance they seem to realize, and certainly openly acknowledge, that a high "heritability" does not imply that the b/w difference in IQ is genetic, but is compatible with any explanation, genetic or environmental, of the difference. But they state that the heritability of IQ is so high (a "mid-estimate" of 60 percent, they claim; we will see why this is wrong) that it is "likely" that the difference is genetic. One starts doubting their good faith or coherence, and the doubts become worse in cases like the following.

They report that since affirmative action was started, there was an important reduction of the b/w difference in scholastic aptitude tests of various types in the last nineteen years. But for them, this fact does not seem to shake their faith that the difference in IQ is genetic. In fact the reduction is so high (30 percent) that, if it continued at the same rate, the gap could become negligible in seventy years. In general, it seems reasonable that in processes of acculturation at least three generations, i.e., a century, is necessary to cancel the initial difference, and that the process may be slower when there is a very strong economic and social difference at the start. Within a century, practically all those who were present at the beginning of the culture change are dead. Moreover, there is always a fraction of the population less exposed to the factors of change, or more resistant to them.

Given the particularly difficult beginnings for people who were in slavery until the last century, and who were only allowed to acquire major civil rights in the 1960s, the reduction observed is very encouraging. Of course, one might think Herrnstein and Murray object to the idea that IQ may have changed, because the computation of the b/w gap reduction is based on measurements of education, not of IQ values, and enthusiasts of

IQ may choose to ignore them as unimportant. But no, they note that the SAT values used for the comparison are highly correlated with the "g" factor (see the next section) and they are therefore all right (but did the IQ gap change? Only by 20 percent or less, they claim). My tendency is the opposite: I prefer measurements that are clearly of acquired knowledge, because this is what you expect from education, and they correspond to something more real than IQ. Moreover, as the history of IQ shows, IQ was initially designed to test the capacity of young children to progress in normal schools in order to be able to recognize early those who would not, and who therefore might profit more from special schools. It is unfortunate that the original aim of IQ tests was completely forgotten, and they were instead employed for the very ambitious aim of "measuring intelligence." This aim is probably unattainable with modern techniques, and there is a culture-bound element in the usual tests that is impossible to recognize and eradicate.

I am convinced that some psychologists and behavioral geneticists (not too many, fortunately) have fallen in love with IQ in a very deep, almost mysterious way. This love affair has become, in many cases, a dedication for life. IQ has a certain number of attractive qualities. It is reasonably reproducible. It is quantitative; some people—arithmophiles—like numbers (I do). It has correlations with many quantities of interest. And what can be more fun than trying to measure that extraordinary quality, intelligence, and relate it to those extraordinary, fashionable things, the genes? Sometimes IQ enthusiasts become even lyrical about it:

"The genes sing a prehistoric song that today should sometimes be resisted but which it would be foolish to ignore"[6].

The trouble with IQ is that it measures a rather small, and perhaps not even the most interesting, part of intelligence. In reality, "intelligence" is very difficult to define and is, in any case, so many sided. The variety of abilities of which the human brain is capable is extremely impressive, and it is important to give recognition to all of them, while IQ specializes in a narrow, and almost dull, form of understanding and capacity to learn; enough to predict if one can profit from ordinary schools, which is important in practice, but not enough to give IQ the magic qualities that its supporters see in it.

The Importance of the Letter g

Like Jensen, Herrnstein and Murray give great prominence to the "g" factor, a mysterious entity which summarizes some general property common to most IQ tests and is, according to IQ enthusiasts, the best available synthesis of "true" intelligence. In fact, g is obtained by current statistical techniques in a way that is simple to execute with available computer programs, but is unfortunately not so easy to explain to the layman, and has no immediate psychological meaning. These statistical methods are similar to those which we have used for the synthesis of gene frequency data and the isolation of patterns hidden in them (see principal components, Chapter 6). The basic mathematical principle behind these methods is highly suitable for the application to the case of gene frequencies, while it is almost certainly a gross approximation when used for IQ tests. This "basic principle" is the use of a "linear" mathematical treatment of the observed data to extract latent patterns contained in them.

I hope the reader will forgive me if I make only a perfunctory attempt to explain this concept: a linear treatment means to use sums of the observed quantities multiplied by appropriate values. These values must be determined by an analysis of the data themselves, using specific conditions based on the problems that one wants to solve. The observed quantities are, in the case of principal components as in Chapter 6, the gene frequencies in the various populations, and in g factor analysis, the numerical results of the various tests in the various individuals. The result of the analysis allows in both cases to transform the original data into many additive components. One of these is more important than all the others, the next is less important, and so on. This provides a simplification because we usually can describe the whole set of data, with little loss, with a few "factors," or "components," depending on the type of analysis we do.

The g factor is analogous to the first principal component, which is by definition the most important component of the variation. In the case of IQ tests, which all have basically a very similar structure, the analysis generates a numerically important g factor, expressing a fraction, by definition large, of the global variation. In the geography of genes the fraction of variation explained by the first principal component is also the most important. It usually is more than 20 percent, sometimes as large as 50 or 60 percent or more. (The reason why in gene geography principal component analysis is particularly appropriate is that the hidden patterns which it dissects out of the data are most frequently due to different, independent population migrations and expansions. These all have a linear effect on gene frequencies, so that linear analysis is a natural mode of separating the various latent patterns. There is no assurance that this is true in the analysis of IQ tests, and in fact there is some indirect evidence that genes do not at all interact linearly one with the other or with the environment in determining IQ or other characters). It is likely that the g factor is barely a statistical artefact, which was given undue prominence by its early discoverers.

Herrnstein and Murray also give great importance to the correlation between "g" and a physical measurement of a "reaction time" suggested by Jensen, which he measures during simple tasks which do not involve conscious thought. He finds reaction time is on average shorter in whites than in blacks. This might be due to a physical characteristic of nerve fibers, but in *The Bell Curve* it is compared with the speed of a computer. It is well known that there has been an increase in the speed of digital computers, which has certainly generated advantages but not qualitative improvements that are due to software innovations. Jensen insisted, especially in his earlier work, in distinguishing a level I and a higher level II of intelligence, which he thought were qualitatively different. He believed blacks could only qualify for the lower level, which would give them a chance of learning only by rote, not by using an unknown, higher type of reasoning reserved to whites (and perhaps also to Asians?). I don't know if the difference in speed is right, or even if it is genetic, but it is very unlikely to generate the qualitative difference Jensen believes.

The Imbroglio of Heritability

We have said it is correct, as Herrnstein and Murray seem to think at the beginning, that the value of the "heritability" estimated within one population, the white one in this case, is irrelevant to the problem of whether the b/w difference is genetic or not. When they later decide that the value of heritability is so high that the b/w difference "is likely to be" or "must be" genetic (they seem to oscillate between these two positions) their contradiction is especially serious, and complicated by the fact that they are wrong in many ways on questions of heritability. They declare the value of heritability to be 60 percent, which is almost twice the correct one, and are so impressed by it that they forget all cautions and go on as if the b/w difference were really genetic. The truth is that even a complete heritability (100 percent) within a population is still perfectly compatible with an entirely non-genetic difference in the means of two populations which do not interbreed; but in any case the heritability of IQ due to genes is closer to 30 percent than to 60 percent.

It is possible that, like many other non-specialists, they have not understood that there are several different ways of defining and measuring heritabilities. These various definitions lead to different quantitative values and have different meanings. I will consider first two categories of heritabilities, called "narrow" and "wide." Narrow values are usually calculated from the similarity of children to parents, or similar quantities; wide heritabilities from the similarity of identical twins. There is a range of mathematical formulas for both types. When discussing differences between mean IQs of different groups it is more appropriate to consider narrow estimates, which are smaller than wide ones. Readers may have heard enough about heritabilities that they may be curious to learn about their nature.

Narrow heritability was developed by breeders for predicting the effect of artificial selection (selection of stocks for special characteristics in domesticated breeds). If you want to select say, for higher yield, in order to produce a commercially more advantageous strain of a domestic plant or animal, you would be interested in predicting in advance, if possible, the success that selection can have. Success is measured as the increase of yield at the end of selection, relative to how much selection was necessary to produce this increase. It depends on the importance of various mechanisms of biological inheritance affecting the characters being selected. "Heritability" of a character is very specifically the measurement which permits the prediction of success in artificial selection. Such experiments usually require many years, and it is obviously important to be able to foretell their results in order to decide whether the effort being planned is worth its cost. But it is narrow heritability, not the wide one, which can tell us about the gain on which one can count in experiments of artificial selection. It may be intuitively clear, perhaps, why narrow heritability is obtained from measurements of the similarity between parents and children. If we take the best 10 percent of cows (best being those which produce, say, the highest amounts of milk), and look at the average amount of milk produced by their children, we can get some idea of the increase in yield we would obtain if we selected for reproduction the ten best cows out of one hundred. Appropriate projections may allow us to predict the results after many generations of continued selection.

One can extend some of these considerations to possible changes of average values of characters under natural selection. Here, however, the situation is more complicated. In principle, differences among human populations in average values of any character, from resistance to diseases to stature to IQ, can be generated by natural selection. The process is not dissimilar from that taking place in artificial selection. Individuals more resistant to some adverse environmental factor, or favored in some other way in the environment in which they live, have greater chances of leaving children, and to pass to them traits which children can inherit. But the process of natural selection is more complicated because for many traits (IQ included) people at the extremes of the range fare less well in natural selection—either survive less or have lower fertility or both—than those having intermediate values of the trait. Natural selection frequently favors central values. All interpretations and predictions of human evolution must take this complication into account. This is an especially difficult time, moreover, because we are witnessing various complicated demographic transitions. Every country or region and every social class has different birth and death rates, which are continuously changing, making demographic and genetic predictions chancy. This somewhat reduces the confidence we can put in narrow heritability for predicting changes which might occur under natural selection; in spite of this, narrow heritability remains the best bet for understanding what might happen under selection, but we cannot reverse the reasoning and explain, on the basis of heritability, differences between the means of two populations due to unknown causes.

Feldman[7] suggested the following thought experiment in order to understand this impossibility: "Take a group of equally white-skinned people from New York. We know that skin color is highly heritable. The group is divided into two subgroups: one subgroup is left to winter in New York, the other goes to Miami. At the end of the winter the skin colors of the groups are compared." Clearly, a difference in skin color will be found between the two groups. In the same way, there are environmental situations which can increase or decrease IQ, leaving genes unchanged (like exposure to the sun in Miami during the winter can darken skin color). The attention paid by many parents to choosing good schools for their children would be meaningless otherwise. Another well-documented environmental effect is that of birth order on IQ, which is also associated to that of family size in IQ: in general, IQ is higher among first born and decreases with increasing birth order (but less so in large families), perhaps because firstborns get the most attention from their parents, and the later born frequently have their older siblings as teachers and not their parents. This teaching activity may help the intellectual development of the older siblings and reduce the learning of the younger ones[8]. Even Herrnstein and Murray, usually inflexible in accepting evidence that IQ has been changed by environmental influences, accept that adoption into good homes can increase average IQ (six points of IQ is a quantity they are willing to accept, but there must be a gradual effect as a function of how "good" a household is!). In fact they seem to waver in their persuasions. One cannot avoid the impression that they oscillate between two extreme views, one of total pessimism, and another in which they are willing to concede that not all ethnic differences are "etched in stone." Perhaps this reflects differences between the two authors.

Geneticists who are familiar with experimental animals and plants are aware of the important effects which environment and random, unaccountable factors often have on the characters they measure. They would hardly consider all differences among different strains of one species, observed in nature and not in controlled environments in the laboratory, as due to genetics. But Herrnstein and Murray chose this strategy, and are coherent with their attitude even when the IQ difference works against whites, as, for example, when comparing East Asians and whites, or against the white majority, as in the comparison of Jews and non-Jews. It seems that for them every difference, great or small, is genetic (apart perhaps, from those due to adoption in good homes!). We have discussed this fact in Chapter 8 and indicated that at least two important environmental factors contribute to the oriental/white difference: the enormous pressure East Asian parents put on their children about their school results, and the fact that East Asians confront a special task, the learning of the Chinese alphabet, which is especially difficult and probably a good training for IQ tests. It is interesting that one IQ test, Raven's progressive matrices, has some however superficial similarity with the problems presented by Chinese characters. Learning to read Chinese may help developing similar skills. In fact, East Asian children seem to do well on Raven's progressive matrices.

We have not yet touched on the major problem with the heritability of IQ, and we will dedicate the following section to it. When Herrnstein and Murray say that the heritability is 60 percent, they imply that the rest, 40 percent, is determined by something else, which is obviously not inherited, or is not genes. This residual set of causal factors is, or was often called, the "environment." It is not so simple, though, because some of these factors are themselves transmitted from parents to children, approximately like genes. It is therefore difficult to distinguish genetic effects on IQ from this other transmission, which we call cultural. We already discussed it in Chapter 8, and whether we like it or not, we have to bring it in by the back door. It has been realized for some time that there might be one way to distinguish them: by studying adopted children, who have both "biological" and adoptive ("cultural") parents, siblings, etc., and comparing their similarities with biological and adoptive relatives. Unfortunately, the procedure is difficult because of the relative rarity of adoptees, and a number of complications that accompany the adoption process. But it is our only hope of separating the effects of the two transmissions, and therefore of avoiding major errors.

The Spoilsport

In Chapter 8 we have cited estimates of heritability of IQ which separated three factors determining IQ: the truly genetic heritability, a cultural heritability derived from cultural transmission across generations, and a residual fraction due to strictly individual factors. We said the three fractions were about equal, i.e, around 33 percent each. A fraction of genetic heritability of 33 percent is a wild cry from the 60 percent accepted by Herrnstein and Murray. I think it is very instructive to give here a history of these estimates, of the problems that were encountered, and how they were solved.

I mentioned before that I fell in love with cultural transmission because I felt a need to understand how cultures as profoundly different from one another as exist on earth can be propagated and maintained with little change over periods likely to be very long. In collaboration with Stanford colleague Marc Feldman we explored the expectations of the various rules by which cultural characters can be transmitted from one individual to another, and their consequences on the maintenance and change of cultures. We published in a 1981 book a general summary of many of our conclusions[9]. I have, in Chapter 8, mentioned the distinction we made then of the two main categories of cultural transmission: vertical (from parents to children) and horizontal (between unrelated people).

It very soon became obvious that vertical cultural transmission and standard genetic transmission can be very similar and determine effects almost indistinguishable from one another. The importance of cultural transmission may be especially great for behavioral traits like IQ, where teaching and learning are important. It therefore may be difficult to separate the two. In other words, it is not impossible that the very high heritability reported for IQ might be due to the approximate summation of the effects of genetic and cultural transmission.

In 1973 we published a paper[10] to show how to calculate the effects of a special type of cultural transmission, "phenotypic transmission." The "phenotype" is the actual realization of a particular character in an individual, for instance the IQ value. In classical genetic theory, the phenotype of an individual is the result of the action of its genes (also called the "genotype" of the individual) in the particular environment in which the individual develops. The environment is usually assumed to vary randomly from one individual to another. Only minor complications were considered in older work, such as the fact that one family might be a better environment than another family for all its members. Our hypothesis was that a parent can give more than his or her genes to a child: it provides a chance to do better (or also worse) either because the child imitates a parent, or reacts against the parent, or is given appropriate teaching by the parent. This direct effect of the phenotype of the parent is an additional transmission, not due to the genes but to the phenotype of the parent, hence the name of phenotypic transmission. To give a simple example, even if it is not strictly IQ, the musical ability of Mozart, although obviously sustained by an exceptionally brilliant musical genotype, undoubtedly gained by the fact that his father was also a very good musician and put considerable effort into teaching him.

It was clear that this type of cultural transmission could make the calculation of heritability, also the narrow one, unacceptable, since the similarity of parents and children, on which narrow heritability is based, is no more only the direct effect of genes, but also of cultural transmission. There would be a "cultural heritability" added to the "genetic heritability," which would inflate the overall similarity between parents and children if the phenotypic transmission of parents acts positively on the value of the trait in the child (if it detracts from it, it would decrease the overall similarity). The heritability calculated from the similarity of parents and children would not reflect in a simple way the action of genes only.

In a later paper[11] we extended the theory to include another way in which cultural transmission can affect a character. Phenotypic transmission—direct teaching or learning—is not the only way in which parents or other relatives can affect a child or, in general, a descendant. The environment in which one lives is also transmitted. Social classes can be highly inherited; caste transmission allows very few or no exceptions. Money and status, friendships, tastes, and many other choices are inherited (culturally). Even if there can be some biological inheritance of say, tastes, or some type of ability to function socially, society has created rules which tend to maintain children in conditions similar to those of their parents. One can therefore have cultural transmission of the "environment" in which one lives: that residual set of causes which, together with genes, determine the phenotype of children and are included in the category of Environment. This set happens to be itself inherited, at least in part. Therefore, both phenotypic transmission and transmission of environment may have to be taken into account. They may be difficult to separate, and it may be simpler to merge them in the category of cultural transmission or inheritance. It remains of paramount importance, however, to distinguish this category from that of genetic influences. It may be added that cultural influences are not restricted to behavioral traits, which are, by definition, developed through social contact and hence in part learned. It is also clear that there will remain factors which influence the phenotype of an individual that are independent from genes or cultural transmission (be it phenotypic or due to transmission of the environment); we call "individual" this category.

Not only IQ, but also physical traits like stature are influenced by transmissible environment, e.g., diet. In Europe at least there has been, and there perhaps still is, a strong stratification of stature by social class, which is unlikely to be genetic. Heritability for stature has been considered extremely high; but it is well known that the average stature has changed enormously in the last two hundred years in almost all the world, and certainly not genetically, but under the influence of environmental changes (in part, at least, diet). Thus also in this physical bulwark, and not only with behavioral traits (which are supposed to be largely learned), cultural inheritance has made an important dent.

There are many data on the similarities of parents and children, of fathers and mothers, of siblings, etc., for IQ, also for adopted children, which permit analysis and estimation of genetic and cultural heritabilities. The mathematics are made more complicated by the necessity to keep track of an important factor: for IQ, and for some conspicuous physical traits (like height), people do not marry at random. On the contrary, they tend to choose individuals more or less very similar to themselves. The reasons for this "assortative mating," however, are not very clear. In part, it is because they choose individuals from the same socioeconomic class—perhaps not consciously but simply because they spend much of their life with individuals of the same class, for instance they go to the same colleges.

A group of geneticists from the University of Hawaii[12] thought they would solve the problem by a statistical technique invented by a famous geneticist, S. Wright, forty-five years earlier. Called "path analysis," it estimates the relative importance of causes ("paths") from complicated diagrams in which causal relationships between the quantities involved are described by connecting them with arrows, specifying the direction of causes. It is clear that genes of parents determine the genes of their children; genes, in combination with environments, determine phenotypes. We thus have arrows connecting the genotype of parents to that of children; arrows from the genotype of an individual and from the environment in which he has developed, going to the phenotype of the individual. If we add phenotypic transmission from parents to children, then there will be arrows from the phenotypes of parents to the phenotype of the child. If we have environmental transmission there is an arrow from the environment of the parents to that of the children. To express the similarity of parents we have a double arrow connecting their phenotypes. Or should we put the arrow between their environments? There is a difference in the consequences. It will depend on whether a husband and wife choose each other on the basis of their phenotype or because of their environment (e.g., their socioeconomic class). Maybe both are important. In general, there are many possible connections between the quantities involved, and every path diagram proposed by the experimenter is a specific hypothesis which must be tested by appropriate techniques. Some quantities like genotypes are unmeasurable, phenotypes (IQ) are; environments are complicated but some aspects of them also can be measured.

The Hawaiian geneticists built a path diagram which, according to their intuition, was the best possible interpretation of the complex system of causes underlying the gene/culture relationships in the inheritance of IQ. Carrying out an analysis of the existing observational IQ data according to the scheme suggested by their path diagram, they came to the conclusion that there was little effect of culture in IQ heritability.

There clearly was a mistake in their theoretical models or calculations, but it was not easy to find. Three mathematically inclined psychiatrists from Washington University School of Medicine in St. Louis, Missouri[13] were able to do so: an arrow had been misplaced by the Hawaiian geneticists in their path diagrams, giving an erroneous representation of the causal system. Putting the arrow at its proper place, it became clear that results were much cleaner. It turned out that cultural and genetic inheritance of IQ had approximately the same importance, giving rise to the figures I mentioned earlier.

One might expect that at this point there would start an endless diatribe between the two groups, but in fact, the whole matter turned out to be an example of the best side of science. The Hawaiian geneticists repeated their calculations and eventually acknowledged that the St. Louis group was right[14]. Correction, and use of a somewhat simpler model, led them to values very similar to those calculated by the St. Louis psychiatrists. The summary of their paper, published in 1982, says: "Rice, Cloninger, and Reich (1980) showed that correlational data on American IQ is consistent with a rather low genetic heritability. Here we confirm their general results with a more parsimonious model. From phenotypic data alone, the estimates of genetic and cultural heritability are 0.31 and 0.42, respectively. Using environmental indices, the parsimonious estimates become 0.34 and 0.26, respectively." One can see this result is very different from their earlier one (1976), in which after an analysis of the same data they had announced that the genetic heritability of IQ remained high (near the level dear to Herrnstein and Murray, 0.60) even though they tried to measure the effect of cultural inheritance, but found it trivial. In our description given in Chapter 8 we have approximated the results they, and the St. Louis group, eventually agreed upon, by saying that each of the three causes, genes, culture, and individual factors, have about the same importance, and therefore each determines about one-third of the total causes.

This was a rather sophisticated quantitative analysis in which three laboratories took part, and that lasted altogether nine years. All papers published in the meantime or afterwards from other groups did not contribute any conclusion as complete or as advanced. Only six other IQ observational data have been published. Their analysis[7] has confirmed the results obtained by Rice and his coworkers. What I have just summarized is the state of the art in the analysis of genetic determination of IQ. It completely confirms our earlier intuition[10] that estimates of genetic heritability of IQ are grossly inflated because of lack of consideration of cultural inheritance.

It is somewhat disconcerting that all these papers are totally ignored in *The Bell Curve*. Unfortunately, they are also ignored by a few other researchers who more recently have collected some new data or performed new analyses with unsatisfactory methods[5]. Researchers who might be called "IQ hereditarians" are in general reporting high heritabilities for IQ without any information on how these calculations have been obtained, or why the other papers here cited have been ignored. It is unlikely that they were not seen or read; they are published in well-known scientific journals. The subject is difficult; it is possible that the IQ enthusiasts who do not cite these seminal papers are aware of them but did not understand them. This is not as incriminating as the possibility that they do not cite them because they are at odds with their own strongly hereditarian conclusions. I hope all these IQ enthusiasts have a chance to read this postscript and take a position with respect to this group of papers. At minimum, they should declare why they continuously ignored them. And they should stop citing genetic heritabilities, like "between 0.60 and 0.70"[15]; "70 percent", unless they also make reference to the sources of these values and why they are different from the lower ones they should use.

The definition of racism we have given in Chapter 9 is based on the persuasion that some races are definitely better than others in some socially important ways, and that the difference is of genetic origin. Herrnstein and Murray seem convinced that IQ is, at least from a socioeconomic point of view, of overriding importance, and, of course, they are convinced it is genetic. They don't seem to be interested in honesty or generosity, but there are no quotients to measure these somewhat out-of-fashion aspects of personality. According to them, if the IQ of certain population classes could be improved, one might hope that poverty, crime, and the other most odious mistakes and failures of modern humanity would disappear or at least be greatly reduced. But they believe that IQ is sufficiently strictly determined by genes and therefore in their views education can have little, if any, effect (although at other times they praise education). I do not know if Herrnstein and Murray are aware that their ideas, if I have represented them correctly, make them racists, and that the same is true of those who share the same ideas. I do not know if they consider it an honor or an offense, or neither. They state that unless affirmative action and social welfare are not greatly attenuated or simply suppressed, anger will become such that "racism will reemerge in a new and more virulent form." Perhaps they distinguish between "good" racism (theirs) and "bad" racism (that of others). That social welfare and affirmative action are often poorly planned is correct, but the result of social laissez-faire would be a disaster, and signify a return to the Hoover era. Among other considerations, my feelings on this issue are profoundly influenced by the effect the great Hoover depression had on my father.

To avoid being misunderstood, let me affirm that my own persuasion is that intelligence is important. I do find it one of the great pleasures of life to spend time with a highly intelligent person. I am also convinced, on the one hand, that there are so many types of intelligence that no single test can measure them, not even those that are socially important. On the other, the IQ view of intelligence is extremely limited and dull. If we consider intelligence to include creativity we may have great difficulties in measuring it, but we acquire marvelous vistas of opportunities and achievements. When we do this, we find that blacks have given so much to the world in terms of art and music, for example, that there must be something wrong in a measurement of IQ that gives them such a low average, or in the concept itself. I also believe that such human qualities as moral stamina, generosity, tenacity, and wisdom for which there do not exist acceptable quotients, are so much more important for society, that the whole subject of IQ should be relegated to those limited applications where it can be truly useful, and not socially disruptive.

References

1. Jensen, A.R., 1969. How much can we boost IQ and scholastic achievement? *Harvard Educational Rev.*

2. Cavalli-Sforza, L. L. and W. F. Bodmer, 1971. *The Genetics of Human Populations.* Freeman and Co.: NY.

3. Bodmer, W. F. and L. L. Cavalli-Sforza, 1970. Intelligence and race. *Scientific American* 223: 19-24.

4. Herrnstein, R. J., 1971. I.Q. *Atlantic Monthly* (September): 43-64.

5. Herrnstein, R. J. and C. Murray, 1994. *The Bell Curve.* The Free Press: NY.

6. Bouchard, T. J., D. T. Lykken, M. McGue, N. L. Segal, and A. Tellegan, 1990. Sources of human psychological differences: The Minnesota study of twins reared apart. *Science* 250:223-228.

7. Feldman, M. W., 1993. *Heritability, Race and Policy.* The Morrison Institute for Population and Resources Studies, Paper No. 0051.

8. Belmont, L. and F. A. Marolla, 1973. Birth order, family size, and intelligence. *Science* 182: 1096-1101.

9. Cavalli-Sforza, L. L. and M. W. Feldman, 1981. *Cultural Transmission and Evolution.* Princeton Univ. Press: Princeton, NJ.

10. Cavalli-Sforza, L. L. and M. W. Feldman, 1973. Cultural versus biological inheritance: phenotypic transmission from parents to children. *Amer. J. Human Genetics* 25: 618-637.

11. Cavalli-Sforza, L. L. and M. W. Feldman, 1978. The evolution of continuous variation. III. Joint transmission of genotype, phenotype and environment. *Genetics* 90: 391-425.

12. Rao, D. C., N. E. Morton, and S. Yee, 1976. Resolution of cultural and biological inheritance by Path Analysis. *Amer. J. Human Genetics* 28:228-242.

13. Rice, J., R. Cloninger, and T. Reich, 1980. Analysis of behavioral traits in the presence of cultural transmission and assortative mating: applications to IQ and SES. *Behavior Genetics* 10:73-92.

14. Rao, D. C., N. E. Morton, J. M. Lalouel, and R. Lew, 1982. Path analysis under generalized assortative mating II. American I.Q. *Genetical Research* 39: 187-198.

15. Jensen, A. R., 1989. Raising IQ without increasing intelligence? *Developmental Review* 9:234-258.

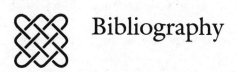 Bibliography

Chapter One: The Oldest Way of Life

Cavalli-Sforza, L., (ed.), 1986. *African Pygmies*. Acad. Press: Orlando, FL.

Hewlett, B. S., 1991. *Intimate Fathers*. University of Michigan Press: Ann Arbor, MI.

Lee, R. B., and I. De Vore, (eds.), 1968. *Man the Hunter*. Aldine: Chicago.

Tobias, P., (ed.), 1978. *The Bushman*. Human and Rousseau: Capetown and Pretoria.

Turnbull, C., 1962. *The Forest People*. Simon and Schuster: New York.

Turnbull, C., 1965. *Wayward Servants*. Nat. Hist. Press: Garden City, NY.

Chapters Two and Three: Human Evolution

Cann, R. L., and M. Wilson, 1987. Mitochondrial DNA and human evolution. *Nature*: 325:31–35.

Coale, A. J., 1974. *The History of the Human Population*: *The Human Population*. A Scientific American Book, W.H. Freeman and Company: San Francisco, p. 15–28.

Foley, R., 1987. *Another Unique Species*: *Patterns in Human Evolutionary Ecology*. Longman: Harlow, Essex.

Isaac, G., 1978. Food Sharing Behavior of Protohuman Hominids. *Scientific American* 238:90–108.

Johannson, D. C., and M. Edey, 1981. *Lucy: The Beginnings of Humankind*. Simon and Schuster: New York.

Jones, S., R. Martin, and D. Pilbeam, (eds.), 1992. *The Cambridge Encyclopedia of Human Evolution*. Cambridge Univ. Press: Cambridge.

Klein, R., 1989. *The Human Career*. U. of Chicago Press: Chicago and London.

Leroi-Gourhan, A., 1982. The Archaeology of Lascaux Cave. *Scientific American* 246:104–112.

Science editorials (on "African Eve"): 255:686–727 (1991): 259:1249 (1993).

Stoneking, M., 1993. DNA and recent human evolution. *Evolutionary Anthropology*: 2:60–73.

Stringer, C. B., 1990. The Emergence of Modern Humans. *Scientific American* 263: 98–104.

Stringer, C., and C. Gamble, 1993. *In search of Neanderthals*. Thames and Hudson: London.

Thorne, A. G., and M. H. Wolpoff, 1992. The multiregional evolution of humans. *Scientific American* 266:28–33.

Trinkaus, E., and W. W. Howells, 1979. The Neanderthals. *Scientific American* 241:118–133.

Vigilant, et al., 1991. African populations and the evolution of human mitochondrial DNA. *Science* 253:1503–1507.

Wilson, A., C., and R. L. Cann, 1992. The Recent African Genesis of Human Genes. *Scientific American* 266:68–73.

Wu, Rukang and Lin Shenglong, 1983. Peking Man. *Scientific American* 248:86–94.

Chapters Four and Five: Human Diversity

Bodmer, W., and L. Cavalli-Sforza, 1976. *Genetics, Evolution and Man*. Freeman and Co.: NY.

Cavalli-Sforza, L., P. Menozzi and A. Piazza, 1994. *History and Geography of Human Genes*. Princeton Univ. Press: Princeton, NJ.

Diamond, J., 1993. *The Rise and Fall of the Third Chimpanzee*. Harper and Perennial: NY.

Jones, S., 1993. *The Language of the Genes*. Harper Collins: London.

Lewontin, R., 1982. *Human Diversity*. Scientific American Library: NY.

Chapter Six: The Last Ten Thousand Years

Ammerman, A. J., and L. Cavalli-Sforza, 1984. *The Neolithic Transition and the Genetics of Populations in Europe*. Princeton Univ. Press: Princeton, NJ.

Bertranpetit, J., and L. Cavalli-Sforza, A genetic reconstruction of the history of the population of Iberian Peninsula. *Annals of Human Genetics* 55:51–67.

Cavalli-Sforza, L., P. Menozzi, and A. Piazza, 1994. *History and Geography of Human Genes*. Princeton Univ. Press: Princeton, NJ.

Cavalli-Sforza, L., P. Menozzi, and A. Piazza, 1993. Demic expansions and Evolution. *Science* 259:639–646.

Menozzi, P., A. Piazza, and L. Cavalli-Sforza, 1978. Synthetic maps of human frequencies in Europeans. *Science* vol. 201, no. 4358: 786–792.

Piazza, A., N. Cappello, E. Olivetti, and S. Rendine, 1988. A genetic history of Italy. *Annals of Human Genetics* 52:203–213.

Sherratt, D., 1981. *The Cambridge Encyclopedia of Archeology* Cambridge University Press: England.

Chapter Seven: The Tower of Babel

Barbujani, G., and R. Sokal, 1990. Zones of sharp genetic change in Europe are also linguistic boundaries. *Proc Natl Acad Sci*, 87:1816–1819.

Bellwood, P., 1991. The Austronesian dispersal and the origin of languages. *Scientific American* 261:70–75.

Cavalli-Sforza, L., 1991. Genes, Peoples and Languages. *Scientific American* 265:104–110.

Cavalli-Sforza, L., A. Piazza, P. Menozzi, and J. Mountain, 1988. Reconstruction of Human Evolution: Bringing together genetics, archeology and linguistics. *Proc. Natl. Acad. Sciences* 85:6002–6006.

Cavalli-Sforza, L., E. Minch, and J. Mountain, 1992. Coevolution of genes and languages revisited. *Proc. Natl. Acad. Sciences* 89:5620–5622.

Crystal, D., 1987. *The Cambridge Encyclopedia of Language*. Cambridge University Press: England.

Gamkrelidze, T., V., and V. V. Ivanov, 1990. The early history of Indo-European Language. *Scientific American* 262:110–116.

Greenberg, J., 1987. *Language in the Americas*. Stanford Univ. Press: California.

Greenberg, J., and M. Ruhlen, 1992. Linguistic origins of Native Americans. *Scientific American* 267:94–99.

Mallory, J. P., 1991. *In search of Indoeuropeans*. Thames and Hudson: London.

Renfrew, 1987. *Archeology and Language*. Cambridge Univ. Press: England.

Renfrew, C., 1987. Origins of Indo-Europeans Languages. *Scientific American* 261:106–114

Ross, P. E., 1991. Trends in Linguistics: Hard words. *Scientific American* 261:70–75.

Ruhlen, M., 1991. *A Guide to the World Languages*. Stanford Univ. Press: California.

Ruhlen, M., 1994. *The Origin of Languages*. John Wiley and Sons: NY.

Wang, W. S-Y., 1989. The Emergence of Language Development and Evolution. Readings from *Scientific American*. W.H. Freeman and Company: NY.

Wright, R., 1991. Quest for the Mother Tongue. *The Atlantic* 267: 39–68.

Chapters Eight, Nine, Ten: Cultural and Genetic Inheritance; Races, Racism, Eugenics, Human Genome

Bodmer, W., and L. Cavalli-Sforza, 1970. Intelligence and Race. *Scientific American* 223:19–29.

Cavalli-Sforza, L., and M. Feldman, 1981. *Cultural Transmission and Evolution*. Princeton Univ. Press: Princeton, NJ.

Cavalli-Sforza, L., 1988. Cultural transmission and adaptation. *International Social Sciences Journal* 116:239–253 (Zagreb Congress Anthropology).

Gould, S. J., 1981. *The Mismeasure of Man*. W. W. Norton: NY.

Kevles, D. J., 1985. *In the Name of Eugenics: Eugenics and the Uses of Human Heredity*. Alfred Knopf: NY.

Kevles, D. J., and L. Hood, (eds.) 1992. *The Code of Codes: Scientific and Social Issues in the Human Genome Project*. Harvard Univ. Press: Cambridge, MA.

Rose, S., R. Lewontin, and L. Kamin, 1984. *Not in our Genes: Biology, Ideology and Human Nature*. Pantheon, NY.

Shipman, P., 1994. *The Evolution of Racism*. Simon and Schuster: NY.

Suzuki, D., and P. Knudtson, 1989. *Genetics: The Clash between Human Genetics and Human Values*. Harvard Univ. Press: Cambridge, MA.

Index

ABO groups, 107–11
Abortion, 246–47, 250–51, 263
Accadic (language), 164
Afar, 30
Africa, 114–15
    African Eve, 62, 66–70, 121
    animals, 31
    australopithecines, 30–31, 39–41
    Bantu expansion, 162
    birth rate, 248
    expansions, 160–62
    farmers, 143
    female circumcision, 217
    genetic tree, 119
    Homo erectus, 41–44, 46, 68
    Homo habilis, 31, 68
    Homo sapiens sapiens, 46, 55, 121
    languages, 9, 18, 175, 179, 185, 201
    Lucy, 30, 32, 38, 41
    marriage, 215
    sub-Saharan, 179
    twentieth century discoveries, 30–32
African Americans, 220, 223, 243–44
African Pygmy, 1–2, 4, 150
    attire, 7
    crossbow, 226
    death, 9
    diet, 6–7
    ethics, 7–8, 21–22
    farming, 15, 144
    future, 22–24
    height, 4, 10–14
    hunting, 23, 203, 226
    language, 9, 18
    lifespan, 8
    marriage and divorce, 9, 150, 215
    net hunting, 2–4
    parenting, 8
    as plantation labor, 14–16
    queens, 150
    social activities, 5
    society, 2
    temperament, 9
    tribes, 5
    tuma, 2
    women, 3, 7, 16, 21, 150, 215
Afro-Asiatic languages, 178–80, 186, 199
Agriculture, 16, 19, 129–38, 262
    development of, 140–43
    and genetics, 144–47
    innovations, 138
    Japan, 163
    map, 141
    and population, 158
    spread of, 135, 153
    tools, 130
AIDS, 252, 254
Ainu, 181
Aka pygmy tribe, 10, 23
Albanian language, 165, 170, 190
Albinos, 100, 205
Aleutian Islands, 185
Alleles, 107
Altaic languages, 175, 180
Altamira caves, 61, 233–34
America, 56, 121–22, 179, 248
American Indians, 109, 112
Amerinds, 110, 114
Amerind languages, 172–75, 179, 182, 185-88, 200

Amino acids, 35–36
Ammermann, Albert, 131, 134, 136, 144, 153
Anatolia, 131–32
Andaman Islands, 20
Andes, 158
Anemia, 80-83, 86
Anglo-Saxon (language), 169, 201
Animals
  cave paintings, 58
  grazing, 155
  hunted by australopithecines, 31
Anteater, 3
Antelopes, 3, 6
Anthropometrics, 114–18
Anthropophagy, 52
Antibodies, 107
Antidotes, 7
Apache languages, 172
Apes, 34, 37–39
Arabia, 158, 161, 178
Arabic language, 179
Aramaic language, 179
Armenian, 190
Arrowheads, 7
Art
  carpets, 132–33
  cave paintings, 58
  fabric decoration, 132
  kilim, 132–33
  wall decoration, 132
Artificial semination, 252–53
Artwork
  aborigine, 22
  caves, 58
  Eskimo, 23
  modern humans, 58
  wall painting, 59
Aryans, 161, 177, 239–41
Asia, 56, 143, 163, 177
Asians, 114, 275
Assortative mating, 278
Assyrian language, 179
Australia, 18–19, 114–15, 117, 159, 179
  agriculture, 143
  language, 19, 179, 186, 194
  modern human fossil, 56
Australopithecines, 30
  aethiopicus, 39
  afarensis, 30, 39, 41
  africanus, 39
  boisei, 39–40

  cranium, 41
  family tree, 40
  map of major finds, 31
  missing links, 39
  robustus, 39
Australopithecus afarensis, 30, 39, 41
Austric language, 186
Austroasiatic language, 185
Austronesian languages, 179
Axe, 44–45
Axum, 161, 199

Bacterium coli, 110
Bamboo, 45
Bananas, 22
Bantu expansion, 162
Bantu farmers, 144
Bantu language, 161
Barbarians, 239
Barley, 131, 142
Basques, 144, 156–57
  language, 165, 170–71, 180, 183, 233–34
  and Rh, 109, 112
Beans, 142
Belgium, 165
Bell Curve, The, 268, 272, 280
Belmont, L., 282
Bengston, J.D., 186
Bertranpetit, Jaume, 234–35
Biological evolution, 57
Biology, and linguistics, 188–91
Birth order, and I.Q., 274–75
Birth rate, 217, 246–48, 267
Bison, 140
Black Americans, 220, 223, 267–69, 281–82
Black, Paul, 189
Blood, 106
  groups, 107–09, 113, 119
  Homo sapiens sapiens, 109–10
  proteins, 119, 124
  Rh, 106, 108–14, 144–46
  samples, 121
  test, 110
  transfusions, 107
  typing, 111
Boas, Franz, 114
Boats, 135
Bodmer, Sir Walter, 220, 258, 267
Body dimensions, 115, 117
Boer farmers, 21

Boganda, Barthelemy, 243
Boisei, 39–40
Bokassa, Jean Bedel, 52, 243
Bones, 28, 53
  carbon dating, 32–34
  crafts, 159
  dragon bones, 29
  teethmarks on, 53
  tools, 57
Boswell, James, 225
Botswana, 21
Bouchard, T.J., 282
Bougainville, 179
Bow, 58
Brachycephalics, 114
Brain, growth of, 102–03
  size, 40, 50, 103, 114, 187
Brazil, 248
Breastfeeding, 134
Brigham, Carl, 227–28
Brittany, 233
Broca, 187
Bronze, 158
Buffalo, 18
Bulgarian tools, 130
Burial, 52-53
Burt, Sir Cyril, 222
Bushmen (San), 18, 23, 166
Buzzati-Traverso, Adriano, 105

Cambodians, 120
Cambridge University, 104, 106, 111
Camels, 158
Camito-Semitic languages, 179
Cancer Research Institute, 258
Cannibalism, 52
Canoe, 158
Cape Town, 166
Carbon, 32
Carbon 14 (dating) method, 32–34
Carpine, Giovanni dal Pian, 20
Caspian Sea, 53
Cassava (manioc), 15, 22
Catal Huyuk, 131–33
Catalan, 165, 193
Caterpillars, 4
Catholicism, 91, 247–48, 250, 263
Cattle, 131, 143
Caucasian languages, 177, 180, 183–85
Caucasoids, 119–20
Cavalli-Sforza, Luca, 113, 119, 153, 160, 197, 224

Cave(s), 239
  Altamira, 61, 233–34
  burial places in, 52–53
  Lascaux, 29, 59, 234
  Le Moustier, 48
  neandertal, 51–52
  paintings, 58
Cell nuclei, 77
Cells, 75
Celtic languages, 171, 177–78, 190
Cephalic index, 114
Cereal crops, 95–96, 131, 136, 142
Chamberlain, Houston Stewart, 240
Chance, 92, 97
Childe, Gordon, 138
Chimpanzees
  catching insects, 7
  and language, 187
  in pygmy diet, 7
China, 142–43, 162, 177, 199–200
  birth rate, 218
  sapiens sapiens relic, 56
Chinese
  abortion, 246–47
  children, 247, 275
  custom, 211
  fossilized bones, 29
  language, 166, 200, 207, 275
  and Rh, 112
Chomsky, Noam, 195
Christianity, 28, 62, 72
Chromosomes, 63, 73, 75–77
Chukchi, 180–81
Cipriani, Lidio, 20
Circumcision, 217
Clicks, (Khoisan), 166, 179
Climate, 56, 114–17, 124, 153, 159
Cloninger, R., 280, 282
Clothes, 60, 159
Clovis Point, 58
Cocoa, 142
Cognates, 171, 180–81
Cold Spring Harbor, 228
Color blindness, 99
Columbus, Christopher, 110, 200
Compass, 158
Conterio, Franco, 100
Contraception, 247
Cook, James, 19
Correlation, 268, 272, 280
Crania, 117
Cranium, 41, 44, 46

Cranium asymmetry, 187
Cro-Magnon, 29, 144
  cave dwellings, 233–35
  and the ice age, 149
  language, 183
  skulls, 50
Crossbow, 226
Cultivators, 16
  intermarriage, 150–51
  language, 160
  polygamy, 216
  pygmy labor and, 14–16
Cultural, 139
  change, 57
  mutations, 214–15
  selection, 191–93
  similarities, 128
  transmission, 207–09, 224, 275–82
Cultural Transmission and Evolution, 224
Cuneiform, 169, 177
Cystic fibrosis, 249

Dance, pygmy, 9–10
Danish, 170
Danube, 136
Dart, Raymond, 30
Darwin, Charles, 28, 34, 44, 87, 116, 188,
    201, 229, 249
Darwinian fitness, 87
Dating
  bones, 33
  carbon 14 method, 32–34
  electron spin resonance (ESR), 55
  Lucy, 32
  radiocarbon, 53
  rock, 33
  stratigraphy, 32
  thermoluminescence, 55
Davenport, C. B., 228
de Gobineau, Count Joseph, 239, 241
de Tocqueville, Alexis, 239
Deafness, 100
Deer, 140
Defense, 158
Demic, 139
Demographic explosion, 129, 133–34
Dene-Caucasian language superfamily,
    183, 186
Denmark, 150
Deoxyribonucleic acid. *See:* DNA
Depression, 253
Dialects, 159, 165, 167

Diasporas, 157
Dictionary of the English Language (1775),
    225
Diet, 6–7, 52, 262
  and height, 114–15, 278
Dinosaur bones, 28
Dipthongs, 193–94
DNA, 30, 34, 75–77, 109, 253, 256, 258
  double helix, 64, 84
  in modern biology, 71
  mitochondrial, 64–65, 67–70, 75, 77
  molecular, 118
  mutation, 65
  and natural selection, 116
  nucleotides, 64, 73, 75, 83–84
  structure of, 62–65, 73
Dolicocephalics, 114
Dominant diseases, 249–50
Dornbusch, Sandy, 224
Double helix, 64, 84
Down syndrome, 93
Dragon bones, 29
Dravidian languages, 161, 177, 180, 200
Drift (language), 194
Duncan, Isadora, 252
Duodenal tumors, 110
Dyen, Isadore, 189–90

Easter Island, 179
Edwards, Anthony, 111, 113
Elamites, 177
Elamitic language, 177
Elbe, 136
Electron spin resonance (ESR), 55
Elephant Man, The, 93
Elephants, 1–3, 203
Embryo, 76
Emigrant farmers, 130
Endogamy, 19, 236
English (language), 168–69, 171, 185,
    188, 193–94, 200, 207
Environmental transmission, 276–80
Enzymes, 119
Epstein-Barr, 257
Epulu pygmies, 6, 11
Eskimo languages, 172, 179–81, 185
Eskimos, 23, 109, 112–13, 179
Estonian, 170
Etcheverry, Michele Angelo, 109
Ethics, 7–8, 21–22
Ethiopia, 30, 161, 178–79, 199
  Lucy, 30, 32, 38, 41

Ethnic cleansing, 240
Ethnography, 25
Eukaryotes, 253
Eugenics, 227
Euphrates, 160
Eurasia, 175, 180
Eurasiatic languages, 174–75, 180–86
Europe, 53–55, 120–23, 149
   expansion in, 137, 153
   genetic dissection, 153
   and the Ice Age, 149
Eutelegenesis, 252
Evolution and progress, 245–46, 262–65
   eugenics, 249–53
   genetic engineering, 253–55
   Human Genome, 255–60
   human genes, 246–48
   humanity, study of, 261–62
   theories and proof, 260–61
Evolutionary tree, 115–20
   linguistics, 189
Evolutionary advantages, 96–97
Expansion, 138, 156, 160–63
   routes, 122
External features, 124–25

Families and superfamilies (linguistic),
   173, 175–77, 180–86
Farmers, 130, 143
   population, 138
Farming, 15, 144
Feldman, Marc, 205, 224, 274, 276, 282
Female circumcision, 217
Fertile crescent, 131
Finnish, 154, 170
Fire, 21, 45
Fired clay materials, dating, 55–56
Fisher, Sir Ronald, A., 104, 106, 111, 220
Fishing tools, 58
Fishing, 140, 159, 163, 199
Fitness, 87
Flint tools, 57
Flower pollen (in neandertal grave), 53
Food, 94–96
Fossils, 27, 29–31
   DNA, 30
   Homo sapiens, 46
   modern human, 56
   volcanic dust and, 30–32
Founder's effect, 98, 109
Frederick the Great, 252

French Society of Linguistics, 183
French, 165, 168, 170–71, 178, 192–93,
   200

Galton, Francis, 249
Gazelles, 3, 6
G factor, 270–72
Gender equations, chromosomes, 73
Genes, 75–77
   and anthropometric features, 116–18
   History and Geography of Human Genes,
   119
   HLA, 146
   human, 246–48
   human genome, 83–84, 255–60
   pseudogenes, 116
   Rhesus gene map, 145
Genetic, 4–5, 144–47, 153
   boundaries, 229
   diseases, 80, 82–83
   distance, 123
   drift, 69, 97–102, 109, 153
   engineering, 253–55
   evolution and linguistic evolution,
   196–98
   geography, Italy, 230–33
   gradient, 150
   markers, 118
   Neolithic Transition and the Genetics of
   European Populations, 144
   purity, 238–39
   screening, 100
   tree, 119
   variation, 20–21, 149
Genome, 83–84, 116, 120, 255–60
Genotype, 277, 279
Genus, 39
German, 165, 168, 170–71, 178, 190,
   195
Germans as supreme race, 239
Germinal cells, 75–77, 79
Gimbutas, M., 155
Glottochronology, 167–68, 171, 180
Goats, 131, 142
Goodall, Jane, 7
Gorillas, in pygmy diet, 7
Gorillas, and language, 187
Grazing, 155
Greek expansion, 156
Greek language, 165, 169–71, 185–86,
   190, 194, 201, 230

Greenberg, Joseph, 169, 171–72, 174–75, 180–81, 183, 198
Grimm laws, 194
Grimm, Jacob, 194
Guide to the World's Languages, A, 174

Hair color, 238
Haldane, J.B.S., 104
Haploid genome, 84
Harappa, 160, 177
Harpoons, 58
Hawaiian Islands, 179
Hebrew language, 179
Height, 4, 10–14, 114–15, 278
Heme, 35
Hemoglobin, 34–35
Herbs, 7
Hereditary diseases, 92–93, 263
Hereditary transmission, 77–79
Heritability, 268–80
Herrnstein, Richard, 221, 268–82
Herodotus, 27
Heterozygote, 86–90, 100
Hewlett, Barry, 226
Hindu, 151
History and Geography of Human Genes, 119
Hitler, Adolf, 239–40, 243
Hittite, 169
HLA genes, 146
Homo, 41
Homo erectus, 41–44, 68
   brain, 44
   major fossil finds, 46
   skulls, 44
   tools, 43
   travel, 44
Homo habilis, 24, 41–42, 68
   cranium asymmetry, 187
   and language, 186–88
   tools, 43
Homo sapiens, 27, 39, 41
   brain, 45
   chronology, 56–57
   differentiation of, 46–48
   family tree, 51
   neanderthalensis, 48, 50–55
   sapiens, 48, 51, 55–56, 68
   sites of major fossil finds, 46
   tools, 47
Homo sapiens sapiens, 46, 55
   birthplace, 121

blood, 109–10
linguistic families, 198
Homozygote, 86–90
Horizontal transmission, 209–12, 276
Horn tools, 57
Horses, 18, 155, 158, 263
Hotelling, Harold, 147
Hottentots (Khoi), 18, 21, 23, 166
Houston, John, 23
Howells, W. W., 50
   head measurements, 116–17
   tree of world populations, 117
Hublin, J. J., 50
HUGO, 258
Human evolution, 44–45
   family tree, 37–39
   See also: Theory of evolution
Human genome, 83–84
Human Genome Diversity, 257–58
Human Genome Project, The, 255–60
Human remains, 28–29
Humans and apes, 34
Humidity, 11–14
Hungarian, 154, 170, 172
Hunter-gatherers, 1–4, 23, 138, 150
   African pygmy, 1–2, 4, 23, 203, 226
   common customs, 21–22
   Indians, North American, 16–18
   meat, 262
   Mesolithics, 136
   and monogamy, 216
   pregnancy, 216
   remaining societies, 18–19
   and women, 3, 7, 16
   world map of, 17
Hunter-nomads, 52
Hunting, 6, 23, 31, 140, 159, 203, 226
Huntington's chorea, 93, 99, 250
Huxley, Thomas, 28, 34, 44
Hypergamy, 151

Iberian peninsula, 165
Ice Age, 149–50
Iceland, 165
Icelandic (language), 165, 167
Identical twins, 85
Illich-Svitych, V. M., 181
Illiteracy, of U.S. immigrants, 227
Impansion, 156
Inbreeding, 19–21
India, 160–61, 248
Indians, North American, 16–17

Indigenism, 139
Indo-European (languages), 160–61, 167, 169, 171–77, 180, 185, 189–90
Indo-Iranian, 240
Indo-Pacific languages, 179, 185–86
Indonesia, 143, 168
Indus Valley, 160, 177
Inequality of Human Races, The, 239
Infant mortality, 246
Intelligence quotient (I.Q.), 219–24, 227, 252, 267–82
Intermarriage, 150–51, 236–37, 244, 248
Inuit, 23
Inventions, 158
Iran, 142, 160–61, 180
Irish, 170, 178
Iron, 158
Iron Age, 150
Isaac, Glynn, 58, 159
Isoglosses, 190–91
Isolation by distance, 190–91
Isotopes, 32
Israel, 53–55, 142, 210
Italian, 165, 167, 169–71, 190, 193
Italy
  contraception and birth rate, 247–48
  genetic geography, 230–33
Ituri, 6, 11, 22, 258
Ivory tools, 57

Japanese, 114, 143, 163
  agriculture, 163
  expansion, 163
  fishing, 163
  I.Q., 223
  language, 180–81, 185
  pottery, 163
Jensen, Arthur, 219–21, 267–70, 272, 282–83
Jews, 99–100, 235–37, 240, 275
Johannsen, Don, 32
Johnson, Dr. Samuel, 225
Jones, Sir William, 169
Judaism, 236

Kamin, Leon, 222
Khoisan, 18, 21, 166, 179, 185–86
Kidd, Judy, 258
Kidd, Kenneth, 258
Kilim, 132–33
Kimura, Motoo, 102, 116, 194

Klasies River, 55
Kluckhohn, Clyde, 206
Korea, 112, 143, 163
  agriculture, 163
  language, 175, 180–81
Kroeber, Alfred, 206
Kruskal, Joseph, B., 189
Kuru, 52
Labov, William, 193
Lactose, 94
Lalouel, J.M., 283
Language, 48, 58, 159, 164, 175–91
  Broca, 187
  change, speed of, 167–68
  cognates, 171, 180–81
  cultural selection, 191–93
  dialects, 159, 165, 167
  drift, 194
  families and superfamilies, 173, 175–77, 180–86
  first appearance of, 186–88
  glottochronology, 167–68, 171, 180
  the great vowel shift, 193
  Guide to the World's Languages, A, 174
  Homo habilis, 186–88
  isoglosses, 190–91
  lexical diffusion, 193–94
  linguistic variations, 166–68
  loan words, 190
  natural selection, 192–93
  Neandertal, 58, 187
  number of, 166–67
  order of words, 168–73
  phonetic laws, 194
  phylum, 173
  proto-languages, 181, 186
  universal roots, 186
  Wernicke, 187
  See also: Specific languages
Lapland, 199
Lapps, 23, 155, 199
Larynx and pharynx, neandertal, 187
Lascaux caves, 29, 233–34
  rock painting, 59
Latin (language), 165, 168–69, 171, 178, 185–87, 192–95, 201
Leakey, Louis, 42
  Rift Valley research, 30
Lees, Robert, 170
Levine, Philip, 107
Lew, R., 283
Lexical diffusion, 193–94
Lexical statistics, 171

Lifespan, 8
Limbs, length of, 114
Linguistics, 153, 159, 164
  and biology, 188–91
  families, 198
  and genetic evolution, 196–98
  variations, 166–68
Linnaeus, Carolus, 39
Livestock, 131, 142, 150, 158, 199, 262
Llama, 158
Loan words, 190
Lorenz, Konrad, 210
Lucy, 30, 38
  cranium, 41
  estimated age, 32
  height, 41
  strata dating, 32
Lumpers, 173
Lykkem, T.D., 282
Lynch, David, 93

Macacus rhesus monkey, 108
Macbeth, 125
Madagascan (language), 168
Madagascar, 143, 179
Maize, 142–43
Malaria, 83, 87, 90–92
Malarial parasite, 125
Male chromosome, 73
Mammoth, 58–59, 140
Mandela, Nelson, 19
Manioc (tapioca), 142
Maps
  Africa, expansions within, 162
  Africa, pygmy groups, 12
  agricultural activity, 141
  hunter-gatherers, 17
  Australopithecines, 31
  Eurasiatic and Nostratic superfamilies, 182
  Europe, Celtic languages, 178
  Homo habilis, 31
  Homo erectus, 46
  Homo sapiens, 46
  Homo sapiens sapiens, expansion routes, 122
  Italy, genetic data analysis (Piazza's), 231
  Italy, isoglosses, 191
  megalithic monuments, 128
  Neandertal settlement sites, 54
  linguistic families (Ruhlen's), 176
  Middle East, fertile crescent, 131

Europe, spread of agriculture in, 135
Europe, Rh frequency, 145–46
Europe, genetic landscape, 149, 154–57
  Middle East, expansion from, 160
  steppe nomads, expansion of, 161
  Na-Dene, Sino-Tibetan, Caucasian
    superfamily, 184
Marolla, E.A., 282
Marriage, 152, 210, 215–16
  and divorce, 9
  endogamy, 236
  intermarriage, 150–51, 236–37, 244, 248
  polygamy, 216
  pygmy, 19, 215
McGue, M., 282
Medicines, pygmy, 7
Megalithic, 127–29
Megaliths, 126–28
Melanesia, 143
  people, 179
Mendel, Gregory, 88, 90, 209–10
Menozzi, Paolo, 119, 145, 148, 153, 174, 197
Mesolithic, 155
Mesolithics, 136, 140
  hunter-gatherers, 150
  and Neolithic peoples, 150–51
Mesopotamia, 263
Mesopotamian languages, 179
Mexico, 143
Microcytemia, 80–83
Micronesians, 179
Middle East, 53–55, 103–04, 131, 138, 142, 146, 149, 157–59, 178
  migration from, 151–53
Middle Stone Age People, 136
Migrationism, 139
Migrations, 120–21, 151–53
Milk, 94
Millet, 142, 162
Mitochondria, 62, 64–65, 67–70, 75, 77
Moccasins, 60
Modern humans, 57–58
Moenjo Daro, 160, 177
Molecular clock, 34, 66
Molecular level DNA, 118
Mongol (language), 181
Mongolia, 175
Mongolism, 92
Mongoloids, 120
Monkey, 6–7, 213

Monogamy, 216
Moroni, Antonio, 101
Morton, Newton, 282–83
Muller, Herman, J., 252
Murder, among pygmies, 7–8
Murray, C., 268–82
Music, pygmy, 9–10
Mutation, 79–83, 85–92, 95–96
  beneficial, 93–95
  of DNA, 65, 68
  fatal, 92
  Y chromosome, 73

Na-Dene languages, 172, 179, 183–85
Namibia, 21
Native American language, 172
Native Americans, 95
Natufian, 142
Natural selection, 92, 116, 192–93, 248,
  273–74
Navajo languages, 172
Nduye mission, 22
Neandertal, 27–29, 49
  bones, 53
  in chronology of Homo sapiens, 56
  flower pollen (in grave), 53
  hunter-nomads, 52
  language, 58, 187
  larynx and pharynx, 187
  last traces of, 55–56
  life of, 51–52
  meat diet, 52
  and modern humans, 57–58
  mousterian tools, 48–49, 52, 57
  necrophagous, 52
  open air sites, 52
  radiocarbon dated, 53
  ritual behavior, 52–53
  skull, 50
  spread of settlements, 53–55
Neanderthalensis, 48, 50–55
  teeth wear, 51
Necrophagy, as cause of kuru, 52
Negative eugenics, 249, 251
Nei, Masatoshi, 119
Neolithic, 130, 142–43
  boat, 135
  cultivators, 160, 177, 233
  expansion in Europe, 137, 153
  farmers, 139, 150
  and Mesolithic peoples, 150–51
  migration from Middle East, 151–53

millet-farming villages, 142
  Rh, 144
  spread of agriculture, 149
Neolithic Transition and the Genetics of
  European Populations, 144
Net hunt, 2–4
Neurofibromatosis, 93
New Guinea, 143
  languages, 179
  stone tools, 143
New Stone Age, 130
New Zealand, 179
Nietzsche, Friedrich, 239
Niger-Kordofanian languages, 179,
  185–86
Nilo-Saharan languages, 179, 185–86
Nitrogen, 33
Nomads, 5
Nongerminal cells, 255
North-south trend, 153–57
Norwegians, 165
Nostratic languages, 180–86
Nucleotides, 64, 73, 75, 83–84, 256–57
Nuraghi, 126–29

Obsidian, 130, 135
Oceania, 120–21, 175
Oceanian aborigines, 121
Oldowan, 42
On The Origin of Species, 189, 201
Onge, 20–21
Opium, Andaman Tribes, 20
Osco-Umbro-Sabellian civilization, 232
Oxen, 61

Pacific islands, 179
Pakistan, 160, 177
Paleolithic migration, 163
Paleontology, 27
Parma River valley, 100–01
Parma University, 101, 145
Path analysis, 278–79
Pauling, Linus, 252
Pavia University, 4, 105, 111, 131, 205,
  232
Peking Man, 44–45
Penicillin, 110
Persecution, 236, 264
Petrification, 29
Phenotype, 277–79
Phenotypic transmission, 277–80
Philippines, 143

Phonetic laws, 194
Phylogenetic tree, 111
Phylum, 173
Piazza, Alberto, 119, 145, 148, 151, 153, 174, 197, 230–31
Pigs, 131
Pingelap Island, 99
Pitcairn Island, 98
Plasmodium falciparum, 90
Plasmodium vivax, 125
Poisons, 7
Poland, 153
Polish, 170
Polo, Marco, 20
Polyandry, 216
Polygamy, 216
Polygyny, 216
Polymorphism, 238
Polynesia, 143
Polynesians, 179, 191
    Austronesian language, 179
    navigators, 179
Population growth rate, 246–48
Portuguese, 165, 200
Positive eugenics, 249, 251
Potassium-argon, 33–34
Potatoes, 142
Pottery, 163
Primitive society, 5
Principal components analysis, 147–50, 152
Proteins, 34–37
Proto-languages, 181, 186
Pseudogenes, 116
Puglia, 127–29
Pumpkin, 142
Pygmies. See Africans
Pyrenees, 58

Qafzeh, 55
Queens, pygmy, 150

Race, racism, and eugenics, 227–29, 267
    biological superiority, 240–41
    European peoples, 233–35
    interracial marriage, 244
    Italy, genetic geography of, 230–33
    Jews, 235–37, 240
    races, number of, 229–30
    racism and racial purity, 237–44, 281–82
Racial persecution, 264

Radiation, and mutation rate, 85
Radioactive, 32–33
Radiocarbon dating, 53
Rao, D.C., 282–83
Raven's progressive matrices, 275
Recessive diseases, 249–50
Reciprocal migrations, 120–21
Reich, Theodore, 280, 282
Reindeer, 23, 199
Religion, 72, 224–26, 265
Religious persecution, 236, 264
Rendine, Sabina, 151
Renfrew, Colin, 160
Rh, 106, 108–14, 144
Rhesus gene map, 145
Rhine, 136
Rice, 162
Rice, J., 280, 282
Rift Valley, 30
Robustus, 39
Rock carvings and drawings, 58
Romania, 165
Romanian, 170–71
Rome, 53
Ruffié, Jacques, 233
Ruhlen, Merritt, 174–77, 185–86, 197
Russia, 146, 163
Russian, 170, 180
Rwanda, 150

Saba, 161, 199
Sail, 158
Salmon, 16
Sanskrit, 169, 177, 194–95
Sapir, E., 194
Sardinia, 126–27
Sarich, William, 37
Sassetti, Filippo, 169
Scandinavia, 199, 211
Scarr, Sandra, 220
Schaaffhausen, Hermann, 28
Schizophrenia, 253
Schleicher, August, 189
Scotland, and language, 178
Scudo, Franco, 202
Scythes, 130
Segal, N.L., 282
Selectively neutral, 116
Selfish DNA, 77, 83
Seminomads, 5
Semitic languages, 179
Sentinel Island, 21

Sephardi Jews, 236
Sex, and sex chromosomes, 73
Sexual selection, 96, 205
Sexual taboo, 216–17
Shaw, George Bernard, 252
Sheep, 131, 142
Shevoroshkin, Vitalj, 182
Shockley, William, 220, 252, 267–69
Siberia, 56, 154, 175
Siblings, 85–86, 221
  twins, 221–23
Simulation, computer, 151–53
Sino-Tibetan languages, 177, 183–85, 200
Skin color, 95–96, 100, 115, 119, 123–25,
    205, 238, 244, 274
Skis, 199
Skulls, 40, 50, 114
Slavic language, 171
Slovenian language, 165
Snow, C.P., 259
Social engineering, 220
Social activities, pygmies, 5
Socioeconomic status, 268, 278, 281
Sociology, 217
Sokal, Robert, 153
Somatic cells, 255
Sorghum, 143
South African Border Caves, 55
Spain, 61, 130, 150
Spanish, 165, 168, 170–71, 200
Spear, 58
Spear-thrower, 60
Spears, 52
Specchie, 127
Species, 39
Spirochaete, 22
Splitters, 173
Stalin, Joseph, 252
Stanford University, 131, 145, 151, 169,
    205, 224, 253, 258, 267, 276
Statistical methods, 147, 271, 278–80
Steppe nomads, 161
Steppes, 155
Sterilization, 249–50, 267
Stern, Curt, 244
Stone
  crafts, 159
  housing, 142
  statuettes, 58
  tools, 42–44, 47–48, 55, 130, 143
Stonehenge, 127
Stratigraphy, 32

Streptococcus, 110
Sub-Saharan, 179
Sudan, 179
Superfamilies, 173, 175–77, 180–86
Surnames, 69–70
Swadesh, Morris, 171
Swedish, 170
Switzerland, 165
Syphilis, 110

Taiwan, 143, 210–11
Taras, 100
Tasmania, 140
Tay-Sachs disease, 99
Teeth, 41, 51
Teeth, shape of, 174
Teethmarks on bones, 53
Tellegan, A., 282
Temperature, and mutation rate, 85
Thalassemia, 80–83, 86, 89–91, 250
Theory of evolution, 72, 74–75
  chance, 92, 97
  evolutionary advantages, 96–97
  field research, 100–01
  genetic diseases, 80, 82–83
  genetic drift, 97–102
  germinal cells, 75–77, 79
  hereditary diseases, 92–93
  hereditary transmission, 77–79
  human genome, 83–84
  migration, 103–04
  mutation, 79–83, 85–92, 95–96
  mutation, beneficial, 93–95
  natural selection, 92
Thermoluminescence, 55
Thorax, circumference of, 114
Thule Eskimos, 19–20
Tibet, 143, 177
Tibetans, 199–200
  marriage, 216
Tigris, 160
Tizard, Barbara, 220
Tocharian A and B (languages), 169
Tomatoes, 142
Tools, 47
  agriculture, 130
  Aurignacian, 48, 57
  Acheulean, 44–45, 48
  ivory, 57
  Mousterian, 48–49, 52, 57
  stone, 42–44, 47–48, 55, 130, 143
Transport, 158

Trombetti, Alfredo, 186
Tuberculosis, 110
Tuma, 2
Turkey, 131–32, 169, 175
Turkish, 181
Turnbull, Colin, 8–9, 11, 22, 204
Twa, 23
Twins, 221–23
Two Cultures, The, 259
Tyson, E., 34

Ugro-Finnic, 154
Ukraine, 153
Ulcers, 110
Universal donor, 107
Universal roots, 186
Upper Paleolithic people, 155
Upper Paleolithic wall painting and
    artifacts, 59–61
Ural mountains, 175–77
Uralic language, 153–54, 170–72, 175,
    180–81, 185
Ussher, James, 27

Vertical transmission, 209–12, 276
Vietnamese, 120
Vikings, 218
Violence, among pygmies, 7–8
Virchow, Rudolf, 28
Virus, 53
Vitamins, 114–15, 124
Volcanic, 30
Vowels and vowel sounds, 193–94

Wagner, Richard, 239–40
Wang, William, 193

Wasserman, 110
Weapons, 158
Weidenreich, Franz, 45
Welsh (language), 168, 170, 178
Wernicke, 187
Westermarck, Edward, 210
Wheat, 131, 142
Wheel, 158
Wilberforce, Bishop Samuel, 28
Wild bee honey, 3
Wilson, Allan, 38, 62, 66-70, 258
Witchcraft, 150–51
Women, 53
    agriculture, 142
    circumcision, 217
    intermarriage, 150
    pygmies, 3, 7, 16, 21, 215
    pygmy queens, 150
    statues and carvings of, 58
Work, 262
World population, 16
    Howell's tree of, 117
Wright, Sewall, 104, 278

Xenophobia, 241
!Xhosa, 18
Xian, 142

Y chromosome, 73
Yale University, 258
Yaws, 22
Yee, S., 282

Zaire pygmies, 22
Zei, Gianna, 231–32
Zhoukoudian (Chou-kou-tien), 44
Zygote, 86–90